WADE DAVIS

SIMON & SCHUSTER PAPERBACKS
New York   London   Toronto   Sydney

The Serpent
and
the Rainbow

SIMON & SCHUSTER PAPERBACKS
Rockefeller Center
1230 Avenue of the Americas
New York, NY 10020

For information about special discounts for bulk purchases,
please contact Simon & Schuster Special Sales:
1-800-456-6798 or business@simonandschuster.com.

Designed by Edith Fowler

Manufactured in the United States of America

20

The Library of Congress has cataloged the hardcover
edition as follows:
Davis, Wade.
   The serpent and the rainbow.
   Bibliography: p.
   Includes index.
   1. Zombiism—Haiti.   2.   Bizango (Cult).
3. Tetrodotoxin—Physiological effect.   4. Datura
stramonium.   5. Davis, Wade, DATE—.   6. Haiti—
Description and travel.   7. Haiti—Religious life
and customs.   8. Haiti—Social life and customs.
9. Pharmacopoeias—Haiti.   I. Title.
BL2530.H3D38   1986   299'.67'097294   85-22114

ISBN-13: 978-0-671-50247-8
ISBN-10:   0-671-50247-6
ISBN-13: 978-0-684-83929-5 (Pbk.)
ISBN-10:   0-684-83929-6 (Pbk.)

Title and part titles based on material in Calligraphic
Alphabets by Arthur Baker, published by Dover Publications, Inc.

To my parents,
to Professor Richard Evans Schultes, who made it possible,
and to John Lennon.

CONTENTS

PART THREE   *The Secret Societies*

He knew the story of King Da, the incarnation of the Serpent, which is the eternal beginning, never ending, who took his pleasure mystically with a queen who was the Rainbow, patroness of the Waters and of all Bringing Forth.

—A. CARPENTIER
The Kingdom of This World

Everything is poison, nothing is poison.

—PARACELSUS

A Note on Orthography

THE ORTHOGRAPHY of the name of the Haitian traditional religion has been the source of some academic debate. The word *voodoo* comes from the Fon language of Dahomey (now Benin) and Togo. It means simply "god" or "spirit." Unfortunately, as a result of the sensational and inaccurate interpretations in the media, Hollywood in particular, the word *voodoo* has come to represent a fantasy of black magic and sorcery. Anthropologists have attempted both to highlight and to avoid this stereotype by using a number of terms including *vodu*, *vodun*, *voudoun*, and *vodoun*. I have followed their lead because I feel, as I hope this book will show, that the rich religion of the Haitian traditional society deserves to be recognized, and what we have come to know as "voodoo" bears little resemblance to it. I use the term *vodoun* because it seems to me to be phonetically the most accurate. However, it is important to note at the outset that the Haitian peasants themselves do not call their religion "vodoun." Theirs is a closed system of belief, and in a world of few alternatives one either "serves the loa"—the spirits—or one does not. Vodoun, from their point of view, refers to a specific event, a dance ritual during which the spirits arrive to mount and possess the believer.

For the sake of clarity, I refer throughout the book to the "vodoun society." This is a concept of convenience, and it also reflects the view of an outsider looking in, not that of a believer surrounded by his spirit realm.

Likewise, the spelling of *zombi* is a matter of some disagreement. *Webster's* prefers *zombie*, the more familiar form, to *zombi*. My Oxford dictionary doesn't even have the term, which reflects the American fascination with Haiti since the Occupation. The sources in the literature are mixed. Seabrook (1929) spelled it *zombie*, as did Deren (1953). Metraux (1972), Huxley (1966), and Leyburn (1941), on the other hand, use *zombi*. Metraux is perhaps the recognized authority on the religion, but to my mind Deren had more intimate contact with the people and is an important source as well—although this has little to do with the spelling of the term.

Of more interest is the derivation. The word probably comes from the Kongo word *nzambi*, which more or less means "spirit of a dead person." This is yet another example of the African roots of the vodoun religion and society.

PART ONE

The Poison

1

The Jaguar

MY FIRST MEETING with the man who would send me on my quest for the Haitian zombi poison occurred on a damp miserable winter's day in late February 1974. I was sitting with my roommate David in a café on a corner of Harvard Square. David was a mountain boy from the West, one generation removed from the family cattle ranch, and just about as rough-cut and restless as Harvard could tolerate. My home was on the rain coast of British Columbia. Both of us had come East to study anthropology, but after two years we had grown tired of just reading about Indians.

A map of the world covered most of one wall of the café, and as I huddled over a cup of coffee I noticed David staring at it intently. He glanced at me, then back at the map, then again at me, only this time with a grin that splayed his beard from ear to ear. Lifting his arm toward the map, he dropped his finger on a piece of land that cut into Hudson's Bay well beyond the Arctic Circle. I looked over at him and felt my own arm rise until it landed me in the middle of the upper Amazon.

David left Cambridge later that week and within a month had moved into an Eskimo settlement on the shore of Rankin Inlet. It would be many months before I saw him again.

For myself, having decided to go to the Amazon, there was only one man to see. Professor Richard Evans Schultes was an almost mythic figure on the campus at that time, and like many other students both within and outside the Department of Anthropology I had a respect for him that bordered on veneration. The last of the great plant explorers in the Victorian tradition, he was for us a hero in a time of few heroes, a man who, having taken a single semester's leave to collect medicinal plants in the northwest Amazon, had disappeared into the rain forest for twelve years.

Later that same afternoon, I slipped quietly onto the fourth floor of Harvard's Botanical Museum. On first sight the Spartan furnishings were disappointing, the herbarium cases too ordered and neat, the secretaries matronly. Then I discovered the laboratory. Most biological labs are sterile places, forests of tubes and flashing lights with preserved specimens issuing smells that could make a fresh flower wilt. This place was extraordinary. Against one wall beside a panoply of Amazonian dance masks was a rack of blowguns and spears. In glass-covered oak cabinets were laid out elegant displays of the world's most common narcotic plants. Bark cloth covered another wall. Scattered about the large room were plant products of every conceivable shape and form—vials of essential oils, specimens of Para rubber, narcotic lianas and fish poisons, mahogany carvings, fiber mats and ropes and dozens of hand-blown glass jars with pickled fruits from the Pacific, fruits that looked like stars. Then I noticed the photographs. In one Schultes stood in a long line of Indian men, his chest decorated with intricate motifs and his gaunt frame wrapped in a grass skirt and draped in bark cloth. In another he was alone, perched like a raptor on the edge of a sandstone massif, peering into a sea of forest. A third captured him against the backdrop of a raging cataract in soiled khakis with a pistol strapped to his waist as he knelt to scrutinize a petroglyph. They were like images out of dreams, difficult to reconcile with the scholarly figure who quietly walked into the laboratory in front of me.

"Yes?" he inquired in a resonant Bostonian accent. Face to face with a legend, I stumbled. Nervously and in a single breath I told him my name, that I came from British Columbia, that I had saved some money working in a logging camp, and that I wanted to go to the

Amazon to collect plants. At that time I knew little about the Amazon and less about plants. I expected him to quiz me. Instead, after gazing for a long time across the room, he peered back at me through his antiquated bifocals, across the stacks and stacks of plant specimens that littered the table between us, and said very simply, "So you want to go to South America and collect plants. When would you like to leave?"

I returned two weeks later for a final meeting, at which time Professor Schultes drew out a series of maps and outlined a number of possible expeditions. Aside from that he offered only two pieces of advice. There was no point buying a heavy pair of boots, he said, because what few snakes I was apt to find generally bite at the neck; a pith helmet, however, was indispensable. Then he suggested enthusiastically that I not return from the Amazon without experimenting with ayahuasca, the vision vine, one of the most potent of hallucinogenic plants. I left his office with the distinct feeling that I was to be very much on my own. A fortnight later I left Cambridge for Colombia without a pith helmet, but with two letters of introduction to a botanical garden in Medellin and enough money to last a year, if spent carefully. I had absolutely no plans, and no perception at the time that my whimsical decision in the café at Harvard Square would mark a major divide in my life.

Three months to the day after leaving Boston, I sat in a dismal cantina in northern Colombia facing an eccentric geographer, an old friend of Professor Schultes's. A week before he had asked me to join him and a British journalist on a walk across "a few miles" of swamp in the northwestern corner of the country. The journalist was Sebastian Snow, an English aristocrat who, having just walked from Tierra del Fuego at the tip of South America, now intended to walk to Alaska. The few miles of swamp referred to was the Darien Gap, 250 roadless miles of rain forest that separated Colombia from Panama. Two years previously a British army platoon led by one of Sebastian's schoolboy friends had traversed the Darien Gap and, despite radio communication, had suffered several casualties, including two unpublicized deaths. Now the intrepid journalist wanted to prove that a small party unencumbered with military gear could do what Snow's schoolmate's military unit could not—traverse the gap safely.

Unfortunately, it was the height of the rainy season, the worst time of year to attempt such an expedition. By then I had some experi-

ence in the rain forest, and when Snow discovered I was a British subject he assumed that I would accompany him all the way. The geographer, on Snow's instructions, was offering me the position of guide and
interpreter. Considering that I had never been anywhere near the Darien Gap, I found the offer curious. Nevertheless I accepted, and gave
the assignment little thought until the night before we were to depart,
when in the clapboard town at the end of the last road before the rain
forest an old peasant woman approached me on the street and offered
an unsolicited appraisal of my situation. My hair was blond, she said,
my skin golden, and my eyes the color of the sea. Before I had a chance
to savor the compliment she added that it was too bad that all these features would be yellow by the time I reached Panama. That same night,
to make matters worse, the geographer, who knew the region far better
than I, somewhat mysteriously dropped out of the expedition.

The first days were among the worst, for we had to traverse the
vast swamps east of the Rio Atrato, and with the river in flood this
meant walking for kilometers at a time in water to our chests. Once
across the Atrato, however, conditions improved, and without much
difficulty we moved from one Choco or Kuna Indian village to the
next, soliciting new guides and obtaining provisions as we went along.
Our serious problems began when we reached the small town of Yavisa,
a miserable hovel that masquerades as the capital of Darien Province,
but is in fact nothing more than a catch basin for all the misfits exiled
from each of the flanking nations.

In those days the Guardia Civil of Panama had explicit instructions
to harass foreigners, and we were the only gringos the Yavisa post was
likely to see for some time. We came expected. Already at a border
post two days west of the frontier an unctuous guard had stolen our
only compass; now at the headquarters we were accused of smuggling
marijuana, an accusation which, however absurd, gave them the excuse
to confiscate our gear. Sebastian became violent and did his best to
prove his maxim that if one yells loudly enough in English, any "foreigner" will understand. This they did not find amusing. Things went
from bad to worse when the sergeant, detailed to rummage through
our gear, discovered Sebastian's money. The mood of the commandant
changed immediately, and with a smile like an open lariat he suggested
that we enjoy the town and return to speak with him in the evening.

We had been walking for two weeks and had hoped to rest in Yavisa for a few days, but our plan changed with a warning we received

that afternoon. After leaving the guardhouse I paddled upriver to an Australian mission post we had heard of, hoping to borrow a compass and perhaps some charts, for the next section of forest was uninhabited. One of the missionaries met me at the dock and acted as if he knew me well. Then, soberly, he explained that according to some of the Kuna at the mission, agents of the commandant intended to intercept our party in the forest and kill us for our money. The missionary, who had lived in the region for some years, took the rumor seriously and urged us to leave as soon as possible. I returned immediately to the jail, discreetly retrieved a few critical items, and then, abandoning the rest of our gear, told the commandant that we had decided to spend a few days at the mission before continuing upriver.

Instead, equipped with two rifles borrowed from the mission, and accompanied by three Kuna guides, we left Yavisa the next day before dawn, downriver.

Our problems began immediately. On the chance that we were being followed, the Kuna led us first up a stone creek bed and then, entering the forest, they deliberately described the most circuitous route possible. Sebastian stumbled, badly twisting an ankle. That first night out we discovered what it meant to sleep on the forest floor at the height of the rainy season. In a vain attempt to keep warm, the three Kuna and I huddled together, taking turns in the middle. Nobody slept. By the end of the second day I had begun to suspect that our riverine Kuna were less than familiar with the forest hinterland, and after three days I realized they were completely disoriented.

Our destination was a construction camp at Santa Fe, which in those years marked the eastern limit of the right-of-way of the Pan-American Highway. A passage that should have taken two days at most stretched on to seven.

When one is lost it is not the absolute number of days that is important, it is the vast uncertainty that consumes every moment. With the rifles we had food, but it never seemed enough, and with the rains each afternoon and night we found little rest. Yet we still had to walk long hours each day through the rain forest, and when one is stripped of all that protects one from nature, the rain forest is an awesome place. Sebastian's injury had not improved, and though he walked courageously he nevertheless slowed our progress. The heat and incessant life seemed to close in, exquisitely beautiful creatures became a plague, and even the shadows of the vegetation, the infinite forms, shapes and

textures, became threatening. In the damp evenings, sitting awake for
long hours while the torrential rains turned the earth to mud, I began
to feel like a crystal of sugar on the tongue of a beast, impatiently
awaiting dissolution.

The worst moment came on the morning of the seventh day. An
hour from our previous night's camp, we stumbled upon the first person we had seen since leaving Yavisa, a solitary and slightly mad woods-
man who had carved a clearing from the forest and begun to plant a
garden. When we asked the direction to Santa Fe, he looked surprised
and, unable to suppress his laughter, he pointed to a barely discernible
trail. At a fast pace, he told us, there was a chance that we could arrive
in another two weeks. His news was so devastating that it was simply
impossible to acknowledge. We had no food left, were physically and
mentally exhausted, and had only enough ammunition left to hunt for
two or three days. Yet we had no choice but to continue, and without
a word passing between us we began to walk, myself in front with one
of the rifles, then Sebastian, followed by the three Kuna. We kept a
fast pace until the forest again held us tangibly, drawing us on into a
hallucinatory passage devoid of will or desire. Into that trance, not
twenty meters before me, leapt a black jaguar. It paused for an instant,
then turned away and took several strides toward what would turn out
to be the direction to Santa Fe, before springing like a shadow into the
vegetation. No one else saw it, but for me it was life itself and, I believe, a portent, for it turned out that Santa Fe was not two weeks
away, but rather a mere two days. That very evening, we found a track
that led to the right-of-way of the highway. It had all been a test of
will, and as we burst into the full sunlight after so many days in the
shade, Sebastian turned to me, placed his arms about my etiolated body,
and said simply that God works in mysterious ways. That evening we
made camp by a clear stream and broiled a wild turkey that one of the
Kuna had shot, and later we slept in the open under a sky full of stars.
For our first night since leaving Yavisa, it did not rain.

The next morning I arose early, certain that I could reach Santa
Fe in a single day. With my belly finally full and the weight of uncertainty lifted from my mind, I basked in the freedom of the open road
and felt an exhilaration I had never known before. My pace increased
and I left the others far behind. The road at first was no more than a
track that curved hypnotically over and around every contour of the
land. Several miles on, however, it rose slowly to the crest of a ridge,

and suddenly from the rise I could see the right-of-way of the Pan-American Highway, a cleared and flattened corridor a hundred meters wide that reached to the horizon. Like a deer on the edge of a clearing, I instinctively fell back, momentarily confounded by so much space. Then I started to walk slowly, tentatively. My senses, which had never been so keen, took in every pulse and movement. No one was behind me, and no one in front, and the forest was reduced to the distant walls of a canyon. Never again would I sense such freedom; I was twenty and felt as if I had reached the heart of where I had dreamed to be.

Before I had left Boston that spring I had advised myself in the frontispiece of my journal to "risk discomfort and solitude for understanding." Now it seemed I had found the means, and my chance meeting with the jaguar remained with me, an affirmation of nature's benevolence. After the Darien Gap I began to look upon Professor Schultes's assignments as koans, enigmatic challenges guaranteed to propel me into places beyond my imaginings. I accepted them easily, reflexively—as a plant takes water. Thus, in time, the Darien expedition was reduced to but an episode in an ethnobotanical apprenticeship that eventually took me throughout much of western South America. I earned my degree in anthropology in 1977 and, following a two-year hiatus from the tropics in northern Canada, returned to Harvard as one of Professor Schultes's graduate students.

Schultes was far more than a catalyst of adventures. His guidance and example gave our expeditions form and substance, while ethnobotany remained the metaphor that lent them utility. He had spent thirteen years in the Amazon because he believed that the Indian knowledge of medicinal plants could offer vital new drugs for the entire world. Forty-five years ago, for example, he had been one of a handful of plant explorers to note the peculiar properties of curare, the Amazonian arrow and dart poisons. Struck by a poison dart, a monkey high in the forest canopy rapidly loses all muscular control and collapses to the forest floor; often it is the fall, not the toxin, that actually kills. Chemical analysis of these arrow poisons yielded D-tubocurarine, a powerful muscle relaxant once used in conjunction with various anesthetics in virtually all surgery. The several species that yielded curare were but a few of the eighteen hundred plants of medical potential identified by Schultes in the northwest Amazon alone. He knew that thousands more remained, elsewhere in the Amazon and around the world. It was to find these that he sent us out. And it was in this spirit that he brought me to the most important assignment of my career.

Late on a Monday afternoon early in 1982 I received a call from Schultes's secretary. I was teaching an undergraduate course with him that semester and expected a discussion of the progress of the class. As I entered his office the venetian blinds were down, and he greeted me without looking up from his desk.

"I've got something for you. Could be intriguing." He handed me the New York address of Dr. Nathan S. Kline, psychiatrist and pioneer in the field of psychopharmacology—the study of the actions of drugs on the mind. It had been Kline who with a handful of others in the 1950s rose to challenge orthodox Freudian psychiatry by suggesting that at least some mental disorders reflected chemical imbalances that could be rectified by drugs. His research led to the development of reserpine, a valuable tranquilizer derived from the Indian snakeroot, a plant that had been used in Vedic medicine for thousands of years. As a direct result of Kline's work, the number of patients at American psychiatric institutions had declined from over half a million in the 1950s to some 120,000 today. The accomplishment had proved a two-edged sword, however, and had made Kline a controversial figure. At least one science reporter had referred to the bag ladies of New York as Kline's babies.

Schultes moved away from his desk to take a call. When he was finished he asked me if I would be able to leave within a fortnight for the Caribbean country of Haiti.

2

"The Frontier of Death"

TWO NIGHTS LATER I was met at the front door of an East Side Manhattan apartment by a tall, strikingly handsome woman whose long hair was pulled up over her head in the manner of a Renoir model.

We shook hands. "Mr. Davis? I'm Marna Anderson, Nate Kline's daughter. Do come in." She turned abruptly and led me through a corridor of clinical whiteness into a large room crowded with color. Approaching me from the head of an immense refectory table was a short man in a white linen suit and an antique vest of silk brocade.

"You must be Wade Davis. Nate Kline. I'm glad you could make it."

There were perhaps nine people in the room, and though Kline made the obligatory introductions, it was in a manner so perfunctory as to let me know that none of them mattered. He lingered only when we reached an elderly man, sitting narrow and stiff in a corner of the room.

"I'd like you to meet one of my oldest colleagues, Professor Heinz

Lehman. Heinz is the former head of psychiatry and psychopharmacology at McGill."

"Ah, Mr. Davis," Lehman said softly, "I am delighted that you have joined our little venture."

"I don't know that I have."

"Yes, well, let us wait and see."

Kline directed me to a sofa where three pleasant but nondescript women sat sipping cocktails, their attentions scattered. A few moments passed in gossip, and then they began to question me about my life with an enthusiasm that made me uncomfortable. As soon as I had a chance I got up and began to circulate, making my way toward the bar, where I poured myself a drink. The room was filled with art—Haitian paintings, antique games and puzzles, a Persian chest with gilt decorations, a small forest of naive early American weather vanes, iron horses poised in flight.

The lights of the city drew me out onto the balcony. Low clouds swept through the high corridors of darkness slowly dissolving the summits of the skyscrapers, and from far below came the sound of tires running over glistening pavement. Looking back through the window, I saw Kline moving vigorously about the room, ushering the last of his dinner guests to the door. His movements seemed ostentatiously virile, reminding me of the kind of elderly man who might ask you in public to place your hand on his chest to measure the strength of his heart. He seemed ill cast as a doctor, exhibiting a vanity more likely in a poet. Lehman, on the other hand—tall, thin, and frail—appeared born to be a psychiatrist, and I couldn't help wondering what vocation he would have pursued had he lived in an earlier age, before men were prepared to yield their feelings to analysis.

Marna joined her father at the door of the apartment, linking her arm lightly in his as they said goodnight to their guests. One sensed immediately the bond between them, how they chose to act as one person, so that his glance became her gesture, which beckoned me in from the balcony.

Emptied of the other guests, the room strangely came to life. Lehman, visibly more at ease, moved to its center. He fixed me with a smile.

"Let me relieve you of any further suspense, Mr. Davis. We understand from Professor Schultes that you are attracted to unusual places. We propose to send you to the frontier of death. If what we are about to tell you is true, as we believe it is, it means that there are men and women dwelling in the continuous present, where the past is dead and the future consists of fear and impossible desires."

I glanced at him skeptically, and then at Kline, who picked up from Lehman automatically.

"The first problem is to know when the dead are truly dead." Kline paused, regarding me deliberately. "Diagnosing death is an age-old problem. Petrarch was nearly buried alive. It haunted the Romans. The writings of Pliny the Elder are full of reports of men rescued at the last minute from the pyre. Eventually to prevent such mishaps the emperor had to fix by law the interval between apparent death and burial at eight days."

"Perhaps we should do the same," interjected Lehman. "Recall the Sheffield case?"

Kline nodded, and turned back to me. "Not fifteen years ago English doctors experimenting with a portable cardiograph at the Sheffield mortuary detected signs of life in a young woman certified dead from a drug overdose."

Lehman added with a smile, "There was an even more sensational case here in New York around the same time. A postmortem operation at the city morgue was disrupted just as the first cut was being made. The patient leapt up and seized the doctor by the throat, who promptly died of shock."

I looked across the table at the two of them, trying to conceal a faint premonition of horror. They were both old, their voices hard and clinical. It was as if the imminent presence of death had so saturated their minds at this late point of their lives that they looked upon it as a source of amusement. I had to remind myself that these men were professionals who had earned some of the highest awards of American science.

"By definition death is the permanent cessation of vital functions." Kline leaned back in his chair, clasping his hands and smiling. "But what constitutes cessation, and how is one to recognize the function in question?"

"Breathing, pulse, body temperature, stiffness . . . whatever," I answered somewhat awkwardly, still uncertain what they were getting at.

"You can't always tell. Breathing can occur with such gentle movements of the diaphragm as to be imperceptible. Besides, the absence of respiration may represent a suspension, not a cessation. As for body temperature, people are pulled out of frozen lakes and snowfields all the time."

"The eyes of the dead tell you nothing," added Lehman. "The muscles of the iris continue to contract for hours after death. Skin color can be useful. . . ."

"Hardly in this case," Kline interrupted, glancing at Lehman. "The pallor of death only shows up in light-skinned individuals. As for heartbeat, any drug that induces hypotension can result in an unreadable pulse. In fact, deep narcosis can manifest every symptom of death: shallow imperceptible breathing, a slow and weakened pulse, a dramatic decrease in body temperature, complete immobility."

Kline poured himself a brandy. " 'No warmth, no breath, shall testify thou livest.' Friar Laurence to Juliet, gentlemen. Perhaps our most famous reference to drug-induced suspended animation."

"In fact," Lehman concluded, "there are only two means of ascertaining death. One is by no means infallible and involves a brain scan and cardiogram. That requires expensive machinery. The other, and the only one that is certain, is putrefaction. And that requires time."

Kline left the room and returned with a document which he presented to me. It was a death certificate in French of one Clairvius Narcisse. It was dated 1962.

"Our problem," Kline explained, "is that this Narcisse is now very much alive and resettled in his village in the Artibonite Valley in central Haiti. He and his family claim he was the victim of a voodoo cult and that immediately following his burial he was taken from his grave as a zombi."

"A zombi. . . ." A dozen conventional questions came to mind, but I said nothing more.

"The living dead," Kline continued. "Voodooists believe that their sorcerers have the power to raise innocent individuals from their graves to sell them as slaves. It is to prevent such a fate that family members may kill the body of the dead a second time, sometimes plunging a knife into the heart of the cadaver, sometimes severing the head in the coffin."

I looked at Kline, then back to Lehman, trying to measure their expressions. They appeared altogether complementary. Kline spoke in visions, in ideas that spun to the edge of reality. Lehman held the reins and balanced the conversation with reason. This made it that much more impressive when he too began to speak of zombis.

"The Narcisse case is not the first to come to our attention. A former student of mine, Lamarque Douyon, is currently the director of the Centre de Psychiatrie et Neurologie in Port-au-Prince. Since 1961, in collaboration with Dr. Douyon, we have been systematically investigating all accounts of zombification. For years we found nothing to them. Then came our breakthrough, in 1979, when our attention was drawn to a series of most singular cases, of which this Narcisse was only one."

The latest, according to Lehman, was a woman, Natagette Joseph, aged about sixty, who was supposedly killed over a land dispute in 1966. In 1980 she was recognized wandering about her home village by the police officer who, fourteen years before, in the absence of a doctor, had pronounced her dead.

Another was a younger woman named Francina Illeus but called "Ti Femme," who was pronounced dead at the age of thirty on February 23, 1976. Before her death she had suffered digestive problems and had been taken to the Saint Michel de l'Attalaye Hospital. Several days after her release she died at home, and her death was verified by a local magistrate. In this case a jealous husband was said to have been responsible. There had been two notable features of Francina's case—her mother found her three years later, recognizing her by a childhood scar she bore on her temple; and later, when her grave was exhumed, her coffin was found to be full of rocks.

Then, in late 1980, Haitian radio reported the discovery near the north coast of the country of a peculiar group of individuals, found wandering aimlessly in what appeared to be a psychotic state. The local peasants identified them as zombis and reported the matter to the local authorities, whereupon the unfortunate party was taken to Cap Haitian, Haiti's second city, and placed under the charge of the military commandant. Aided in part by an extensive media campaign, the army had managed to return most of the reputed zombis to their home villages, far from where the group had been found.

"These three instances," Lehman remarked, "while curious, were still no more substantial than many others that had periodically surfaced in the Haitian press."

"What made the Narcisse case unique," said Kline, "was the fact that he happened to die at an American-directed philanthropic institution which, among its many features, keeps precise and accurate records." Thus Kline began to describe the extraordinary case of Clairvius Narcisse.

In the spring of 1962, a Haitian peasant aged about forty approached the emergency entrance of the Albert Schweitzer Hospital at Deschapelles in the Artibonite Valley. He was admitted under the name Clairvius Narcisse at 9:45 P.M. on April 30, complaining of fever, body ache, and general malaise; he had also begun to spit blood. His condition deteriorated rapidly, and at 1:15 P.M. on May 2 he was pronounced dead by two attendant physicians, one of them an American. His sister Angelina Narcisse was present at his bedside and immediately notified the family. Shortly after Narcisse's demise an elder sister, Marie Claire, arrived and witnessed the body, affixing her thumbprint to the official death certificate. The body was placed in cold storage for twenty hours, then taken for burial. At 10:00 A.M., May 3, 1962, Clairvius Narcisse was buried in a small cemetery north of his village of l'Estère, and ten days later a heavy concrete memorial slab was placed over the grave by his family.

In 1980, eighteen years later, a man walked into the l'Estère marketplace and approached Angelina Narcisse. He introduced himself by a boyhood nickname of the deceased brother, a name that only intimate family members knew and that had not been used since the siblings were children. The man claimed to be Clairvius and stated that he had been made a zombi by his brother because of a land dispute. In Haiti, the official Napoleonic code states that land must be divided among male offspring. According to Narcisse, he had refused to sell off his part of the inheritance, and his brother had, in a fit of anger, contracted out his zombification. Immediately following his resurrection from the grave he was beaten and bound, then led away by a team of men to the north of the country where, for two years, he worked as a slave with other zombis. Eventually the zombi master was killed and the zombis, free from whatever force kept them bound to him, dispersed. Narcisse spent the next sixteen years wandering about the country, fearful of the vengeful brother. It was only after hearing of his brother's death that he dared return to his village.

The Narcisse case generated considerable publicity within Haiti and drew the attention of the BBC, which arrived in 1981 to film a short documentary based on his story. Douyon, meanwhile, had considered various ways to test the truth of Narcisse's claim. To exhume the grave would have proved little. If the man was an impostor, he or his conspirators could well have removed the bones. On the other hand, had Narcisse actually been taken from the grave as a zombi, those responsible might have substituted another body, by then impossible to identify. Instead, working directly with family members, Douyon designed a series of detailed questions concerning Narcisse's childhood—questions that not even a close boyhood friend could have answered. These the man claiming to be Narcisse answered correctly. And over two hundred residents of l'Estère were certain that Narcisse had returned to the living. By the time the BBC arrived Douyon himself was convinced. To close the circle, the BBC took a copy of the death certificate to Scotland Yard, and there specialists verified that the fingerprint belonged to the sister, Marie Claire.

It was several moments before I could accept the seriousness of their conclusion. I stood up and moved, escaping the white whorls of cigarette smoke, anxious to shake loose a dozen thoughts and questions.

"How do you know this isn't an elaborate fraud?"

"Perpetrated by whom and for what end?" Kline replied. "In Haiti a zombi is a complete outcast. Would a leper stand upon Hyde Park Corner and boast of his disease?"

"So you are saying that this Narcisse was buried alive."

"Yes, unless you believe in magic."

"What about oxygen in the coffin?"

"His survival would have depended on his level of metabolic activity. There is a medically documented case of an Indian fakir consciously reducing his oxygen consumption and surviving ten hours in an airtight box hardly larger than a coffin."

"It is worth pointing out," interjected Lehman, "that damage due to oxygen deprivation would be progressive."

"In what sense?"

"If certain brain cells are without oxygen for even a few seconds they die and can never recover their function, for as I probably don't have to tell you, there is no regeneration of brain tissue. The more primitive parts of the brain, those that control vital functions, can endure greater abuse. Under certain circumstances the individual may lose personality, or that part of the brain that deals with thought and voluntary movement, and yet survive as a vegetable because the vital centers are intact."

"Precisely the Haitian definition of a zombi," noted Kline. "A body without character, without will."

Still incredulous, I turned to Kline.

"Are you suggesting that brain damage creates a zombi?"

"Not at all, at least not directly. After all, Narcisse was pronounced dead. There must be a material explanation, and we think it is a drug."

Finally I knew what they wanted from me.

"I first came across rumors of a zombi poison some thirty years ago," said Kline. "During my first years in Haiti I tried unsuccessfully to obtain a sample. I did meet an old voodoo priest who assured me that the poison was sprinkled across the threshold of the intended victim's doorway and absorbed through the skin of the feet. He claimed that at the resurrection ceremony the victim was administered a second drug as an antidote. Now both the BBC and Douyon have sent us very similar reports."

"Douyon brought us a sample of a reputed zombi poison some months ago," said Lehman. "We tested it on rats but it proved to be completely inert. However, a brown powder given to us recently by one of the correspondents of the BBC may be of greater interest. We prepared an emulsion and applied it to the abdomen of rhesus monkeys; it caused a pronounced reduction in activity. We have absolutely no idea what the powder was made from."

Lehman's grave dark face had changed; it was luminous, trembling. I found his excitement contagious. Yes, it was completely conceivable that a drug might exist which, if administered in proper dosage, would lower the metabolic state of the victim to such a level that he would be considered dead. In fact, however, the victim would remain alive, and an antidote properly administered could then restore him at the appropriate time. The medical potential of such a drug could be enormous . . . as Kline obviously appreciated.

"Take surgery," he said. "Someone is about to have an operation. What do they want to be sure of?" Before I could reply, he said, "Their surgeon? They want to know that the surgeon is qualified, but the truth is that most surgery is absolutely routine. The real liability, the hidden danger that kills hundreds of patients every year, no one even thinks about."

Lehman was restless, anxious to finish Kline's thought, but Kline went on. "Anesthesia. Every time someone goes under, it is an experiment in applied pharmacology. The anesthesiologist has his formulae and his preferred chemicals, but he combines them on the spot, depending on the type of operation and the condition of the patient. Each case is unique and experimental."

"And hazardous," Lehman added. Kline held his empty brandy glass to the light.

"We cloak all uncomfortable truths in euphemism," Kline said, moving back to the table toward me. "General anesthesia is essential, often unavoidable, always dangerous. That makes everyone, especially the physicians, uncomfortable. Hence we joke about getting knocked out, as if it were a straightforward procedure. Well, I suppose it is. Bringing someone back undamaged, however, is not."

Kline paused. "If we could find a new drug which made the patient utterly insensible to pain, and paralyzed, and another which harmlessly returned him to normal consciousness, it could revolutionize modern surgery."

It was my turn to interrupt. "And make somebody a lot of money."

"For the sake of medical science," Lehman insisted. "That's why it behooves us to investigate all reports of potential anesthetic agents. We must have a close look at this reputed zombi poison, if it exists."

Kline moved across the room like a man at odds with something more than himself. "Anesthesis is only the beginning. NASA once asked me to consider the possible application of psychoactive drugs in the space program. They would never admit it, but basically they were concerned with how they were going to keep the restless astronauts occupied during extended interplanetary missions. This zombi poison could provide a fascinating model for experiments in artificial hibernation."

Lehman looked at Kline impatiently. "What we want from you, Mr. Davis, is the formula of the poison." The bluntness of his statement, however expected, pushed me back from the table, and I turned my back on them both, stepping toward a sliding glass door, until I felt myself caught like a fly in the cross mesh of their gaze.

I turned back to them. "What about contacts?"

"We will be touch with Douyon. And perhaps you should call the BBC and speak with their correspondent."

"That's it?"

"That's all we know."

"And my expenses?"

"We have a small fund put aside. Just send us the bills."

There was nothing more to ask. They were like two major currents, Kline torrid and surging, Lehman passive and subdued; they had come together, determined to act. My assignment as outlined succinctly by Kline was to travel to Haiti, find the voodoo sorcerers responsible, and obtain samples of the poison and antidote, observing their preparation and if possible documenting their use.

As I went out the apartment door, Kline handed me a sealed manila envelope, and it was then I realized that they had assumed all along that I would take the assignment. I didn't look back, even as I heard their voices continuing behind me.

Kline's daughter Marna caught up with me in the lobby. It was late, and I walked her back to her Sixty-ninth Street studio. Outside on the streets a thin drizzle had turned the pavement to pools of yellow light. The storm had passed and the city once again carried its own sounds. Marna hadn't said anything during the meeting, and she didn't speak now. I asked her about a photograph I had noticed in the apartment, of a frail white-haired man sitting at a desk, reaching a hand across a pair of ivory-handled revolvers.

"François Duvalier. Eugene Smith took it when he and my father were in Haiti."

"Your father knew Papa Doc?"

She nodded.

"How?"

"When they set up that institute where Douyon has the zombis. The one named after him."

"After Duvalier?"

"No," she said with a laugh, "my father. He's been going to Haiti for twenty-five years."

"I know. Ever go with him?"

"Yes, all the time, but . . ."

"Like it?"

"Sure, it's wonderful. But listen, you ought to understand something. He really believes zombis exist."

"You don't."

"That's not the point."

Outside her apartment house, an empty cab approached and I hailed it. We said goodnight. It was hours past the last air shuttle, so I directed the cabbie to Grand Central Station and waited for the night train to Boston. Once on board, I opened the envelope that Kline had given me. Besides money and an airplane ticket there was one Polaroid photograph, a dull image of a poor black peasant, whom a note identified as Clairvius Narcisse. I found myself cradling his face in my hand, astonished how a mere photograph could make the exotic seem intimate. I still held it as the train pulled out, and then finally I glanced at the airline ticket. I had one week to try to piece together a biological explanation that would fit the limited data.

3

The Calabar Hypothesis

I ENJOY TRAINS, and in South America whenever possible I rode them, sitting on the open ends, savoring the waves of tropical scents that the passage of the train whipped into an irresistible melange. By comparison to those creaking Latin caravans, so alive in human sweat and wet wool, smelling of a dozen species of crushed flowers, American trains are sadly sterile, with a heavy atmosphere that makes the air taste used. Still, the rhythm of the rails is always seductive, and the passing frames race by like so many childhood fantasies, alive in color and light.

But leaving New York, it was not to the train that I owed my strange sense of release. I questioned my reflection in the train window, puzzled by a range of inexpressible feelings and ideas. "The frontier of death"—it was that phrase of Lehman's that haunted me most, pulling me back from the borders of sleep, leaving me alone in an empty train car measuring the passing night by the periodic shuffling of the conductor's feet.

Kline and Lehman. I weighed their words, groping for hidden

meanings or clues, but kept returning to the bare facts of the case. These did not tell me much, but they were enough to get started, and, moreover, they mercifully grounded my imagination.

A poison sprinkled across a threshold was presumably absorbed through the feet. If true, this implied that its principal chemical constituents had to be topically active. From descriptions of the wandering zombis, it appeared likely that the drug induced a prolonged psychotic state, while the initial dose had to be capable of causing a deathlike stupor. Since in all likelihood the poison was organically derived, its source had to be a plant or animal currently found in Haiti. Finally, whatever this substance might prove to be, it had to be extraordinarily potent.

Knowing very little about animal venoms, I reviewed the toxic and psychoactive plants I had become familiar with during my six-year association with the Botanical Museum. I thought of plants that could kill, and others that could lead one past the edge of consciousness. There was only one that even nominally met the criteria of the zombi poison. It was also the one plant that during all my investigations, and through all my travels, I had dared not imbibe—a hallucinogenic plant so dangerous that even Schultes, for all his stoic experimentation, had never sampled. It is a plant that has been called the drug of choice of poisoners, criminals, and black magicians throughout the world. Its name is *datura*, "the holy flower of the North Star."

My tired thoughts broke into fragments that landed on a distant night, cold and clear as glass, in the high Andes of Peru. A brown dusty trail curved past agave swollen in bud and rose to an open veranda flanked on three sides by the adobe walls of the farmhouse. Against one wall sat the patient, alone and strangely solemn. He had been a prosperous fisherman a season ago, before the currents shifted and the warm tropical waters came south to strangle the sea life of the entire coast. As if conforming to some bitter law of physics, his personal life had mimicked the natural disorder: his child had taken ill, and then his wife fled with a lover. In the wake of these events the poor man disappeared from his village, only to reappear a month later, a simulacrum of death, naked and quite insane.

For two weeks the *curandero* had sought in vain to divine the source of such misfortune. With his inherent eye for the sacred he had laid out the power objects of his altar—stone crystals, jaguar teeth, murex shells, whale bones, and ancient huacas that rose methodically to

touch an arc of colonial swords impaling the earth. In nocturnal ceremonies he and the patient had together inhaled a decoction of alcohol
and tobacco from scallop shells carefully balanced beneath each nos-
tril. Invoking the names of Atahualpa and all the ancient Peruvian kings,
the spirits of the mountains and the holy herbs, they had imbibed
achuma, the sacred cactus of the four winds. The curandero's son had
led the madman on mule on a slow passage high into the mountains to
bathe in their spiritual source, the lakes of Las Huaringas. All to no
avail. The visions had come, only weak and incomprehensible, and even
the pilgrimage to the healing waters had done little to free the deranged
man from his stubborn misery.

It was left for the curandero to work alone, to seek a solution in
a stronger source, in some supernatural realm that might break a normal man. It was a solitary task, and leaving his patient sitting alone, he
slipped away, walking with a stoop, sheltered by a worn poncho and
an enormous hat that covered all of his face save his chin, which protruded like the toe of an old boot. He would engage a different set of
visions—confusing, disorienting, unpleasant—and he would approach
them not as a man of knowledge who might interpret and manipulate
his spirit world, but rather as a supplicant who in just touching the
realm of madness unleashed by the plant might attain revelation. It was
a frightful prospect to relinquish all control, to lose all sense of time
and space and memory. But he had no choice, and he approached his
task with resignation, like the bearer of an incurable disease.

He retreated into a small stone hut, sealed by a broken door that
turned his movements into vertical slices of light. I peered through
these cracks at his shadowy figure moving in purposeful, increasingly
smaller circles, the way a dog does before it beds down for the night.
Once on the ground, he removed his hat, revealing a vaguely distorted
face—distended blue-black lips and an elephantine nose that drooped
precariously toward his mouth. The flesh had collapsed on his cheekbones, his eyes were lost in shadow. He sat quietly, accepting but not
acknowledging the ministrations of his assistant, who carefully arranged a bed, a large basin of water, and a small enamel bowl of dark
liquid. The assistant came out and took his place discreetly to one side
of the door. He beckoned me to join him, and I moved close. We remained still, peering into the dark room, our breathing silenced by the
light wind falling on the tin roof.

The curandero clasped the enamel bowl as a rural priest might
hold a chalice, with his whole hands, firmly and without grace. Nod-

ding first to the four corners of the hut, he drank slowly, deliberately, wincing slightly only once before draining the vessel. Then he sat profoundly still, with the calm that invariably follows such irrevocable acts.

The potion took effect quickly. Within half an hour he had sunk into a heavy stupor, his eyes fixed vacantly on the ground, his mouth sealed shut, his face suddenly bloated and red. His nostrils flared, and several minutes later his eyes began to roll, foam issued from his mouth, and his entire body shook with horrible convulsions. He plunged deeper and deeper into delirium, breathing spasmodically, kneading the earth with his long bony fingers like a cat exploring for fissures that might release him from his madness. Agonizing screams sliced into the night. He attempted to stand, only to fall and lie flat on the ground, thrashing the air with his arms. Suddenly he lunged for the basin of water, like a man whose skin is aflame or whose throat is parched. Then with a final anguished spasm, he collapsed and lay still.

This was *cimora*, the tree of the evil eagle, the closest botanical relative of datura.

The pale lavender light of dawn shone through the opaque membrane of the train window, a mirage of life, and finally, the slow scuffle of feet in South Station led me out into the morning light and onto the streets of Boston.

The city was just coming awake, and I felt far too agitated to sleep. I got to the Botanical Museum by the time it opened and had to wade through a horde of schoolchildren being dragged by a schoolmaster to the exhibits before I could climb the iron staircase and finally reach the private library on the upper floor.

The air was musty, the usual reassuring scent. From behind the oak cabinet that held the ancient folios and the original editions of Linnaeus, I extracted some leather-bound monographs, seeking impatiently that first clue that might solidify my intuitions. I found it in an old brown-paged catalog written some forty years ago. Datura did grow in Haiti, three species, all of them introduced from the Old World. I scanned the list of common names, names that frequently reflect popular applications of the plant. One of the species was *Datura stramonium*. To the Haitians this was *concombre zombi*—the zombi's cucumber! With a quiet sense of satisfaction I retreated to my favorite chair, and within minutes fell fast asleep.

●

A key touched the outer lock. Professor Schultes walked in, his arm cradling several volumes.

"Don't you usually sleep in your office?" he asked wryly. We exchanged pleasantries, and then I briefed him on the zombi investigation.

Schultes shared my instincts for datura, and together we spent the morning building the case.

There was no question that species of datura are topically active. Sorcerers among the Yaqui Indians of northern Mexico anoint their genitals, legs, and feet with a salve based on crushed datura leaves and thus experience the sensation of flight. Schultes felt that quite possibly the Yaqui had acquired this practice from the Spaniards, for throughout medieval Europe witches commonly rubbed their bodies with hallucinogenic ointments made from belladonna, mandrake, and henbane, all relatives of datura. In fact, much of the behavior associated with the witches is as readily attributable to these drugs as to any spiritual communion with the diabolic. A particularly efficient means of self-administering the drug for women is through the moist tissues of the vagina; the witch's broomstick or staff was considered a most effective applicator. (Our own popular image of the haggard woman on a broomstick comes from the medieval belief that witches rode their staffs each midnight to the sabbat, the orgiastic assembly of demons and sorcerers. In fact, it now appears that their journey was not through space, but across the hallucinatory landscape of their minds.)

That the plant is capable of inducing stupor is suggested in the origins of the name itself, which is derived from the *dhatureas*, bands of thieves in ancient India that used it to drug their intended victims. In the sixteenth century the Portuguese explorer Christoval Acosta found that Hindu prostitutes were so adept at using the seeds of the plant that they gave it in doses corresponding to the number of hours they wished their poor victims to remain unconscious. A later traveler to the Indies, Johann Albert de Mandelslo, noted in the mid-seventeenth century that the women, closely watched by their husbands yet tormented by their passion for the novel Europeans, drugged their mates with datura and then "prosecuted their delights, even in the presence of their husbands," who sat utterly stupefied with their eyes wide open. A more macabre use was recorded from the New World, where the Chibcha Indians of highland Colombia administered a close relative of datura to the wives and slaves of dead kings, before burying them alive with their deceased masters.

The pharmacological evidence was solid. Datura was topically active, and in relatively modest dosage induced maddening hallucinations and delusions, followed by confusion, disorientation, and amnesia. Excessive doses resulted in stupor and death.

Yet I had another intuitive reason to implicate datura in the zombi phenomenon. Among many Amerindian groups, life is conceptually divided into stages beginning with birth and progressing through initiation, marriage, and finally death. The transition from one stage to the next is often marked by important ritualistic activity. When I had first heard Kline and Lehman describe Narcisse's account of his resurrection from the grave, it had struck me as a kind of passage rite—a perverse inversion of the natural process of life and death. Perhaps more than any other drug, datura is associated with such transitional moments of passage, of initiation and death. The Luisena Indians of southern California, for example, felt that all youths had to undergo datura narcosis during their puberty rites in order to become men. The Algonquin and other tribes of northeastern North America also employed datura, calling it *wysoccan*. At puberty, adolescent males were confined in special longhouses and for two or three weeks ate nothing but the drug. During the course of their extended intoxication, the youths forgot what it was to be a boy and learned what it meant to be a man. In South America the Jivaro, or Shuar—the famed headhunters of eastern Ecuador—give a potion called *maikua* to young boys when at the age of six they must seek their souls. If the boy is fortunate, his soul will appear to him in the form of a large pair of creatures, often animals such as jaguars or anacondas. Later the soul will enter the body.

For many Indian tribes datura is closely associated with death. In parts of highland Peru it is called *huaca*, the Quechua name for grave, because of the belief that those intoxicated with the plant are able to divine the location of the tombs of their ancestors. The Zuni of the American Southwest chew datura during rain ceremonies, often placing the powdered roots in their eyes as they beseech the spirits of the dead to intercede with the gods for rain. Perhaps more than any other clue, it was this connection between datura and the forces of death and darkness that had offered the first indication of the makeup of the zombi poison.

Our attention naturally focused on *Datura stramonium*, the species known in Haiti as the zombi's cucumber. Although this plant appears to have been native to Asia, its value as a drug was such that it was

widely dispersed throughout Europe and Africa long before the time
of Columbus. Because Schultes knew of no reports of the indigenous use
of this species by Caribbean Indians, and because of the African origins of the Haitian, I was particularly curious about its distribution in
West Africa.

Later that day, it came as no surprise to read that many tribes
made use of Datura stramonium. The Hausas of Nigeria used the seeds
to heighten the intoxication of ritual beverages. It was given to Fulani
youths to excite them in the sharo contest, the ordeal of manhood.
Witch doctors in Togo administered a drink of its leaves and the root
of a potent fish poison (Lonchocarpus capassa) to disputants who appeared before them for a settlement. In many parts of West Africa
the use of Datura stramonium in criminal poisonings still takes a unique
form: women breed beetles and feed them on a species of the plant,
and in turn use the feces to kill unfaithful lovers.

If the poison originated in Africa, it was reasonable to assume that
the antidote that Kline had referred to would be found there as well.
It was therefore with some satisfaction that I discovered that the recognized medical antidote for datura poisoning is derived from a West
African plant. The substance is physostigmine, a drug first isolated
from the Calabar bean (Physostigmine venenosum), a climbing liana
that grows in swampy coastal areas of West Africa from Sierra Leone
south and east as far as the Cameroons. It is especially well known on
the Calabar coast near the Gulf of Guinea at the mouth of the Niger
River, precisely the region from which many of the forefathers of the
Haitian people embarked as slaves for the plantations of the New
World.

A brief sojourn in the ethnographic literature revealed that the
eighteenth-century French plantation owners of Saint Domingue, now
Haiti, chose their slaves with some care. The carnage on the plantations
was horrendous, and as they found it cheaper to bring in adult Africans
than to raise slaves from birth, prodigious numbers had to be imported.
In a mere twelve years, for example, between 1779 and 1790, the slave
ships that plied the coast of Africa from Sierra Leone to Mozambique
unloaded close to four hundred thousand slaves in Saint Domingue.
Although these unfortunate individuals came from virtually every
corner of the continent, the plantation owners clearly had certain
preferences. The Senegalese were highly regarded for their superior
morality and taciturn character—an ironic assessment for the slavers to

make—whereas the people from Sierra Leone and the Ivory and Gold Coasts were considered a stubborn group likely to revolt and desert. The Ibos of the southern Slave Coast of what is now Nigeria worked well but were prone to suicide. The people from the Congo and Angola were highly regarded, and large numbers of them were imported. But it was the peoples bought along the Slave Coast that were preferred above all others, and much of the European slave trade concentrated there. Such was the scale of the trade that in the Kingdom of Dahomey it became a national industry, with the economy of the entire country based on annual expeditions against neighboring peoples. Many of the captured victims—Nagos, Mahis, and Aradas of the western Yoruba, among others—were taken down the Niger, and there they fell into the hands of a notorious and opportunistic tribe of traders, the Efik of Old Calabar.

Originally fishermen, the Efik were ideally situated near the estuary of the Niger River to take advantage of the bitter competition for slaves. The prevailing winds and currents forced all ships returning to Europe or the Americas from the Ivory and Gold Coasts to pass eastward toward the Slave Coast and close to the shores of Efik lands. As avaricious middlemen soon equipped with European arms, the Efik came to control the entire trade with the hinterland; their name, in fact, is derived from an Ibibio-Efik word meaning "oppress," a name received from those neighboring tribes on the lower Calabar and Cross rivers whom the Efik prevented from establishing direct contact with the white traders.

The Efik were no more cooperative with the Europeans. They demanded lines of credit and were regularly entrusted with trade items—salt, cotton cloth, iron, brass, and copper valued in the thousands of pounds sterling. In addition to exchanging goods for slaves, the Europeans had to pay a duty for the privilege of trading with the Efik chiefs. Though some of the slave ships anchored off the coast for up to a year, no European was ever permitted to touch the shore; they could only wait, sometimes for months, to pick up their cargo.

Apparently each major Efik settlement was ruled outwardly by an *obong*, or chief, who enforced laws, mediated in disputes, and led the armed forces in times of war. Besides this secular authority, however, there was a second and perhaps more powerful social and political force, a secret society called the *Egbo*, or leopard society. The Egbo was a male hierarchical association consisting of several ranks, each of which had a distinctive costume. Although the Egbo and the secular authority were institutionally separate, in practice powerful individuals of the tribe sat in council for both groups. Fear of the clandestine and mysterious Egbo was often exploited by members of the secret society themselves, and the obong invariably ranked high in the society.

Under a secret council of community elders that constituted the supreme judicial authority, the leopard society promulgated and enforced laws, judged important cases, recovered debts, and protected the property of its members. It enforced its laws with a broad range of sanctions. It could impose fines, prevent an individual from trading, impound property, and arrest, detain, or incarcerate offenders. Serious cases resulted in execution, by either decapitation or fatal mutilation— the victim was tied to a tree with his lower jaw sliced off.

The tribunal of the secret society determined guilt or innocence by a judgment of a most singular form. The accused was made to drink a toxic potion made from eight seeds of the Calabar bean, ground and mixed in water. In such a dose, physostigmine acts as a powerful sedative of the spinal cord, and causes progressively ascending paralysis from the feet to the waist, and eventual collapse of all muscular control, leading to death by asphyxiation. The defendant, after swallowing the poison, was ordered to stand still before a judicial gathering until the effects of the poison became noticeable. Then he was ordered to walk toward a line drawn on the ground ten feet away. If the accused was lucky enough to vomit and regurgitate the poison, he was judged innocent and allowed to depart unharmed. If he did not vomit, yet managed to reach the line, he was also deemed innocent, and quickly given a concoction of excrement mixed with water which had been used to wash the external genitalia of a female.

Most often, however, given the toxicity of the Calabar bean, the accused died a ghastly death. The body was racked with terrible convulsions, mucus flowed from the nose, the mouth shook horribly. If a person died from the ordeal, the executioner gouged out his eyes and cast the naked body into the forest.

At any one time during the later years of the slave trade, the Efik lands were crowded with newly acquired slaves, most of them thoroughly demoralized. To keep order and discipline, the Efik depended on the agents and executioners of the Egbo. During the weeks and sometimes months that the slaves were held awaiting shipment to the

Americas, they must have heard of the gruesome ordeal of the Calabar bean, and many may have undergone the judgment themselves.

Here was an exciting possibility. Datura was a violently psychoactive plant well known and widely used in Africa as a stuporific poison by at least some of the peoples that had been exported to Haiti. The Calabar bean, which yields the recognized medical antidote for datura poisoning, came from the same region, and knowledge of its toxicity would almost certainly have passed across the Atlantic. African species of datura were apparently common throughout contemporary Haiti. The Calabar bean, though unreported from Haiti, has a hard outer seed coat and could have easily survived the transoceanic passage, and like datura it could have been later sown deliberately or accidentally, in the fertile soils of Saint Domingue.

Thus after a day in the library I had something concrete, a hypothesis that, however tenuous, at least fitted the sparse facts of the case. Knowledge of the pharmacological properties of these two toxic African plants presumably traveled across the Atlantic with the slaves to Saint Domingue. Then, adapted to fit new needs, or perhaps conserved as adjuncts to ancient magical practices, they provided the material basis upon which the contemporary belief in zombis was founded. My Calabar hypothesis was only conjecture, but at least it was a beginning, a skeletal framework on which other ideas and new information could be draped, until a solution to this extraordinary mystery took form.

The hypothesis was simple and elegant but, as it would turn out, quite wrong. Yet in pursuing it I had unwittingly uncovered a sociological connection that would eventually prove a key to the entire zombi phenomenon—the secret societies of the Efik.

4

White Darkness
and the Living Dead

I TRAVELED to Haiti in April 1982, armed only with my tentative hypothesis, Kline's introduction to Lamarque Douyon, the Port-au-Prince psychiatrist who had in his clinic Clairvius Narcisse, and two names I received from the BBC in London: Max Beauvoir, described as a sophisticated member of the Haitian intellectual elite and a noted authority on the vodoun religion; and Marcel Pierre, the vodoun priest, or *houngan,* from whom the BBC had obtained their sample of the reputed poison, a man one of their correspondents had labeled the "incarnation of evil."

As my plane approached Haiti, it was not hard to understand why Columbus had responded as he had when asked by Isabella to describe the island of Hispaniola. He had taken the nearest piece of paper, crumpled it in his hand, and thrown it on the table. "That," he had said, "is Hispaniola."

Columbus had come to Haiti by way of the island of San Salvador, his first landfall in the Americas, where the natives had told enticing tales of a mountainous island where the rivers ran yellow with soft stones. The admiral found his gold, but more excitedly he discovered a tropical paradise. In rapture he wrote back to his queen that nowhere under the sun were there lands of such fertility, so void of pestilence, where the rivers were countless and the trees reached into the heavens. The native Arawaks he praised as generous and good, and he beseeched her to take them under her protection. This she did. The Spaniards introduced all the elements of sixteenth-century civilization, and as a result within fifteen years the native population was reduced from approximately half a million to sixty thousand. European rapacity carved away the forest as well, and with the disappearance of the rich stands of lignum vitae and mahogany, rosewood, and pine, the delicate tropical soils turned to dust and blackened the rivers. Now, looking out of the plane at the barren slopes and the dry, desolate landscape, I saw the European arrival on that island four hundred years before like the coming of a plague of locusts.

The capital city of Port-au-Prince lies prostrate across a low, hot tropical plain at the head of a bay flanked on both sides by soaring mountains. Behind these mountains rise others, creating an illusion of space that absorbs Haiti's multitudes and softens the country's harshest statistic: a land mass of only ten thousand square miles inhabited by six million people. Port-au-Prince is a sprawling muddle of a city, on first encounter a carnival of civic chaos. A waterfront shantytown damp with laundry. Half-finished, leprous public monuments. Streets lined with *flamboyant* and the stench of fish and sweat, excrement and ash. Dazzling government buildings and a presidential palace so white that it doesn't seem real. There are the cries and moans of the marketplace, the din of untuned engines, the reek of diesel fumes. It presents all the squalor and grace of any Third World capital, yet as I drove into the city for the first time, I noticed something else. The people on the street didn't walk; they flowed, exuding pride. Physically they were beautiful. They seemed gay, careless, jaunty. Washed clean by the afternoon rain, the whole city had a rakish charm. And it wasn't just how things appeared, it was something in the air, something electric—a raw elemental energy I had never felt elsewhere in the Americas. Yet while I sensed this feeling immediately, and remained aware of it constantly during all the months I was to spend in the country, I would not understand it for some time. That first day coming in from the airport, however, I did receive a clue—a sight I would see many times again in Haiti. There in the late afternoon sun was a single individual, quite sane and very happy, standing alone, dancing with his own shadow.

I checked into the Hotel Ollofson, a filigree mansion draped in bougainvillea and saturated with the air of that long-forgotten era when the United States occupied Haiti. Leaving my bags at the desk, I left immediately for the home of Max Beauvoir, in Mariani, south of the capital beyond a frenetic thoroughfare known as the Carrefour Road. All the traffic that drains the hinterland to the south passes here, but it is less a route than a happening, a condensation in a few kilometers of all the life and drama of the city. My driver treated it as a free-for-all, flying recklessly past stevedores bent double beneath loads of ice and charcoal, fish and furniture. From every direction kaleidoscopic "tap taps," the Haitian buses, overflowing with nattering passengers and their goods, careened on and off the roadway in search of still more cargo. In front of shops, bossy *marchand* ladies flaunted their wares, artisans carved shoes out of tires, or forged carriage parts from iron bars. And everywhere Dominican girls in tight-fitting rayon hung heavy in doorways lined with caladium. It is a dirty, lively, gaudy boulevard where the roadside houses climb atop one another vying for the attention of those passing by.

Just beyond a cemetery of whitewashed tombs, the road bursts into the open, and for the first time since entering the Carrefour one senses the sea at hand. About three kilometers beyond, where the nose of the mountain touches the water, my driver turned into a grove of trees.

A porter met me at the gate, and I followed him through a marvelous garden toward a small outbuilding on the edge of the property. There, among a dusty collection of amulets and African art, Max Beauvoir awaited me. He was immediately impressive—tall, debonaire in dress and manner, and handsome. Fluent in several languages, he questioned me at length about my previous work, my academic background, and my intentions in Haiti. In turn, I offered my initial hypothesis concerning the use of poisons in the creation of zombis.

"And should you find these zombis? Will you not laugh at their misery?"

"I can't say. Maybe, just as I laugh at my own."

He smiled. "Spoken like a Haitian. Yes, we do laugh at our misfortune, but we reserve that right for ourselves." He hesitated, pulling deeply on a cigarette. "I am afraid you shall be looking for this poison for some time, Mr. Davis. It is not a poison that makes a zombi, it is the *bokor*."

"The bokor?"

"The priest who serves with the left hand," he said cryptically. "But that is a false distinction." He paused. "In a way, we are all bokors, we houngan. The houngan must know evil to combat it, the bokor must embrace good in order to subvert it. It is all one. The bokor who knows the magic can make anyone a zombi—a Haitian living abroad, a foreigner. Likewise, I can treat a victim, should I choose. It is our force, and our greatest defense. But this talk is in vain. This is a land where things are not the way they seem."

Beauvoir led me back to my car, exchanging pleasantries but offering neither comment nor information related to my assignment. Instead, having revealed to me that he was a vodoun priest, a houngan, he asked me to return that night to witness his vodoun ceremony. The land we stood on was his *hounfour*—his temple, a sanctuary, and shrine.

Max Beauvoir held a commercial vodoun ceremony every night. Anyone was invited, and there was a ten-dollar charge for tourists that supported his family and the thirty or more people who worked for him. I arrived around ten and was taken into the peristyle, the roofed court of the hounfour, and was led around a semicircle of tables to the one where Beauvoir sat at the head. A waiter brought me a drink. Beauvoir invited me to scan the assembly of visitors, an eclectic gathering that included some French sailors, several groups of Haitians, an anthropology professor from Milan who had visited earlier in the day, a pair of journalists, and a party of American missionaries. There was the sound of a rattle, and Beauvoir directed my attention to the rear of the temple.

A white-robed girl—one of the *hounsis*, or initiates in the temple—came out of the darkness into the peristyle, spun in two directions, then placed a candle on the ground and lit it. The *mambo*, or vodoun priestess, repeated her motion bearing a clay jar, then carefully traced a cabalistic design on the earth, using cornmeal taken from the jar. This, Beauvoir explained, was a *vévé*, the symbol of the *loa*, or spirit, being invoked. The mambo next presented a container of water to the cardinal points, then poured libations to the centerpost of the peristyle, the axis along which the spirits were to enter. Further libations were offered to each of the three drums and the entrance of the temple. Then, with a flourish, the mambo led the initiates into the peristyle and around the centerpost, the *poteau mitan*, in a counterclockwise direction until they knelt as one before the houngan. Bearing a rattle, or *asson*, Beauvoir led the prayer, an elaborate litany that invoked in hierarchical order the spirits of the vodoun pantheon. He recited in an ancient ritual language whose sounds evoked all the mysteries of an ancient tradition.

Then the drums started, first the penetrating staccato cry of the *cata*, the smallest, whipped by a pair of long thin sticks. The rolling rhythm of the second, the *seconde*, followed, and then came the sound of thunder rising, as if the belly of the earth were about to burst. This was the *maman*, largest of the three. Each drum had its own rhythm, its own pitch, yet there was a stunning unity to their sound that swept over the senses. The mambo's voice sliced through the night, and against the rising chords of her invocation the drummers beat a continuous battery of sound, a resonance so powerful and directed it had the very palm trees above swaying in sympathy.

The initiates responded, swinging about the peristyle as one body linked by a single pulse. Each hounsis remained anonymous, focused inward and turned away from the audience toward the poteau mitan and the drums. Their dance was not a ritual of poised grace, of allegory; it was a frontal assault on the forces of nature. Physically, it was a dance of shoulders and arms, of feet flat on the ground repeating deceptively simple steps over and over. But it was also a dance of purpose and resolution, of solidity and permanence.

For forty minutes the dance went on, and then it happened. The maman broke—fled from the fixed rhythm of the other two drums, then rushed back with a highly syncopated, broken counterpoint. The effect was one of excruciating emptiness, a moment of hopeless vulnerability. An initiate froze. The drum pounded relentlessly, deep solid blows that seemed to strike directly to the woman's spine. She cringed with each beat. Then, with one foot fixed to the earth like a root, she began to spin in a spasmodic pirouette, out of which she soon broke to hurtle about the peristyle, stumbling, falling, grasping, thrashing the air with her arms, momentarily regaining her center only to be driven on by the incessant beat. And upon this wave of sound, the spirit arrived. The woman's violence ceased; slowly she lifted her face to the sky. She had been mounted by the divine horseman; she had become the spirit. The loa, the spirit that the ceremony had been invoking, had arrived.

Never in the course of my travels in the Amazon had I witnessed a phenomenon as raw or powerful as the spectacle of vodoun possession that followed. The initiate, a diminutive woman, tore about the peristyle, lifting large men off the ground to swing them about like children. She grabbed a glass and tore into it with her teeth, swallowing small bits and spitting the rest onto the ground. At one point the mambo brought her a live dove; this the hounsis sacrificed by breaking its wings, then tearing the neck apart with her teeth. Apparently the spirits could be greedy, for soon two other hounsis were possessed, and for an extraordinary thirty minutes the peristyle was utter pandemonium, with the mambo racing about, spraying rum and libations of water and clairin, directing the spirits with the rhythm of her asson. The drums beat ceaselessly. Then, as suddenly as the spirits had arrived, they left, and one by one the hounsis that had been possessed collapsed deep within themselves. As the others carried their exhausted bodies back into the temple, I glanced at Beauvoir, and then back across the tables of guests. Some began nervously to applaud, others looked confused and uncertain.

It was only the beginning of an extraordinary night. More was to follow, Beauvoir explained. What we had just seen were the rites of *Rada*, derived almost directly from the services of the deities of Dahomey. In Haiti, the Rada have come to represent the emotional stability and warmth of Africa, the hearth of the nation. Customarily in the Port-au-Prince region they are followed by those of a new nation of spirits, forged directly in the steel and blood of the colonial era. These are the *Petro*, and they reflect all the rage, violence and delirium that threw off the shackles of slavery. The drums, dancing, and rhythm of their beat are completely distinct. Whereas the Rada drumming and dancing are on beat, the Petro are offbeat, sharp, and unforgiving, like the crack of a rawhide whip.

The spirits arrived again, only this time riding a fire burning at the base of the poteau mitan. The hounsis was mounted violently—her entire body shaking, her muscles flexed—and a single spasm wriggled up her spine. She knelt before the fire, calling out in some ancient tongue. Then she stood up and began to whirl, describing smaller and smaller circles that carried her like a top around the poteau mitan and dropped her, still spinning, onto the fire. She remained there for an impossibly long time, and then in a single bound that sent embers and ash throughout the peristyle, she leapt away. Landing squarely on both feet, she stared back at the fire and screeched like a raven. Then she embraced the coals. She grabbed a burning faggot with each hand, slapped them together, and released one. The other she began to lick, with broad lascivious strokes of her tongue, and then she ate the fire, taking a red-hot coal the size of a small apple between her lips. Then, once more she began to spin. She went around the poteau mitan three times until finally she collapsed into the arms of the mambo. The ember was still in her mouth.

After the ceremony ended, a number of the audience came over to speak with Beauvoir, but I was drawn toward the fire at the foot of the poteau mitan. I felt its heat. I teased an ember out of the flames, and lifted it between two pieces of kindling.

"It surprises you."

I turned to the voice and found one of the hounsis, her white dress still wet with sweat.

"Yes, it is amazing."

"The loa are strong. Fire cannot harm them."

With that, she excused herself and moved toward Beauvoir's table. Then I realized she had spoken perfect English. This was Rachel Beauvoir. She was sixteen, and she walked as if her dancing never stopped.

It seemed like days later when I returned to the Ollofson that evening. The hotel appeared to have shifted its mood yet again. In the daylight when I had arrived it was a white palace, fragile and pretty, a gingerbread fantasy of turrets and towers, cupolas and wooden minarets decorated in lace, which paint alone kept from collapsing into the sea. By late afternoon it had fallen into desuetude, its beams swollen by the moist heat, its atmosphere dense from the impending storm. Later, in the wake of the deluge that tumbled every day like an avalanche onto the tropical plain of the city, the building's facade washed clean, it glowed again with warmth and beauty in the soft air of dusk. Now, by night and a shrouded moon, it had grown morbid, abandoned, overgrown, staring out over the city with shuttered windows, its gates bound by lianas, its gardens unkempt and wild.

I sat on the veranda, too restless to sleep, attempting to make sense out of what I had seen at Beauvoir's. There was no escaping the fact that a woman in an apparent state of trance had carried a burning coal in her mouth for three minutes with impunity. Perhaps even more impressive, she did it every night on schedule. I thought of other societies where believers affirm their faith by exposing themselves to fire. In São Paulo, Brazil, hundreds of Japanese celebrate the Buddha's birthday by walking across beds of coals, the temperature of which has been measured at 650 degrees Fahrenheit. In Greece, tourists regularly watch the firewalkers at the village of Ayia Eleni, acolytes who believe that the presence of Saint Constantine protects them. The same sort of thing goes on in Singapore and throughout the Far East. Western scientists have gone to almost absurd lengths to explain such feats. Generally they invoke the "Leidenfrost point," citing the effect that makes drops of water dance on a skillet. This theory suggests that just as heat vaporizes the bottom of the water droplet as it approaches the skillet, a thin protective layer of vapor is formed between the burning rocks, for example, and the firewalkers' feet. I had to smile as I recalled this explanation. To my mind it begged the question entirely. After all, a water droplet on a skillet is not a foot on a red-hot coal, nor lips wrapped about an ember. I still burn my wet tongue if I place the lit end of a cigarette on it. And my own experience in Indian sweat lodges, where the temperatures may reach the boiling point, had taught me that only concentration and the guidance of the medicine man allowed one to endure such a test. Now, after what I had seen at Beauvoir's, any explanation that did not take into account the play of mind and consciousness, belief and faith, seemed hollow. The woman had clearly entered some kind of spirit realm. But what impressed me the most was the ease with which she did so. I had no experience or knowledge that would allow me either to rationalize or to escape what I had seen.

"And you, *mon cher*, what are you here for?" The words startled me, and I turned to face a narrow man dressed in fine linen, perched on the edge of the hotel veranda like a shorebird. In his right hand spun an ebony cane inlaid with silver.

"A journalist, no doubt. And which of the many faces of this land shall you see? Shall you see the misery, the suffering, and call it the truth?"

He took three slow steps across the veranda and dropped gracefully into a wicker chair, crossing his legs as he sat. Above him the slow whirl of a wooden fan paced his practiced words like a metronome. He seemed fraudulent, yet I was drawn to him as one is to a caricature. He turned despondent.

"My country, my beautiful country, is run by fools. Watch them descend from the heights in their silver cars, hands clasped to teak steering wheels. *Mon cher*, they smile like satyrs that have deflowered a nation."

He spoke almost like a drunkard, yet his eyes were clear.

"Perhaps you shall know the other Haiti, if you can bear it. We are a nation of three—the rich, the poor, and myself. We have all forgotten how to weep. Our wretched past is forgotten as a foul dream, an awkward interlude."

I stood up to leave.

"I see I frighten you. My deepest apologies."

I bade the stranger goodnight and crossed the veranda toward my room. He watched me in a faded mirror.

I awakened early the next day and decided to drive north to the town of Saint Marc to look up Marcel Pierre, the houngan who had provided the BBC with its sample of the reputed zombi poison. Beauvoir called me before I could leave, and when I told him my plans he suggested that I take his daughter Rachel with me as an interpreter.

I was waiting on the veranda of the Ollofson when they arrived. Rachel wore a cotton dress, and as they walked up the alabaster steps of the hotel, the patterns ran together like a watercolor.

The trip up the coast was unlike any other drive I had taken in the Americas. It began by the docks, where the black shanties face the cruise ships, and men with legs like anvils drag rickety carts laden with bloody cowhides. Passing out of the city through the lush canefields of the Cul de Sac Plain, it reached the slopes of the Chaine de Matheux and turned back to the sea. Further on, among the wattle-and-daub houses thatched in palm, the concrete ancestral tombs, and the long lines of sleek bodies and bicycles by the roadside, one sensed Africa at hand. All the produce of this surprisingly abundant land is carried on the head—baskets of eggplant and greens, bundles of firewood, tables, a coffin, a single piece of cane, sacks of charcoal, buckets of water, and countless unidentifiable drab bundles. Everything large or small is carried atop out of habit as much as necessity, like a delightful but defiant challenge to the laws of gravity. By the roadside in the shaded tunnel formed by planted neem trees, the passages of rural life come on theatrical display.

I felt lucky to have Rachel with me. Like a child set loose in a carnival, she delighted in the landscape and took pleasure in pointing out things I could not have seen, let alone understood—all the incidental visual anecdotes that for her somehow formed a whole. Yet she herself seemed such a mix of lives. The night before I had sat at the peristyle transfixed by the magic of her and the other hounsis; now as we spoke—and we did continuously—I realized that she was also a high school senior, and an American one at that. Rachel, in fact, had been born in the United States and had spent the first ten years of her life in Massachusetts. Her family, once back in Haiti, and with an eye to her future university education, had enrolled her in the private school run for all the English-speaking children of the foreign diplomatic community. Hence, as we traveled north to try to obtain a poison from Marcel Pierre, our conversation ran from zombis to high school yearbooks, proms, college admission boards, and back to the loa. I don't know if she noticed how strange it all seemed to me. Perhaps she was thinking the same thing.

"What do you do for a living?" she asked at one point.

"Mainly, I'm an ethnobotanist."

"What's that?"

"Somewhere between an anthropologist and a biologist. We try to find new medicine from plants."

"Have you found any?"

"No." I laughed.

"I'd like to study anthropology, or literature. But I think I like anthropology better."

"So do I. You get tired of just books."

"I already am." She was looking out the windshield watching the slopes of the mountains that loomed above the coastal road. "Somewhere around here a friend of my father saw a ball of fire come out a cave," she said.

Saint Marc was still. In the white heat of midday, not even a dog ventured forth. The pallid luster of haze shimmered at the ends of long dusty streets. Buildings of brick and wood, ravaged by time, stood braced by the surrounding hills, scabrous and stripped of vegetation, and mile after mile of featureless hills rose to distant mountains trapped by the horizon.

Rachel's aunt had been the mayor of Saint Marc, but we didn't need her help to locate Marcel Pierre. He was well known. A mechanic pointed out a bar and dry goods store at the northern end of town in an area known as the Wasp's Gate. Marcel Pierre owned the Eagle Bar, as the place was called. Behind it he had built his hounfour.

In the shadowy doorway of the bar a Dominican woman leaned

on a Wurlitzer, apparently inured to the raucous music that even at this time of day poured from the machine. Rachel greeted her. The woman motioned us to a table on the porch, then turned and left, trailing the sour scent of cheap perfume. The music was unbearable, we agreed, and we got up to venture inside. The rear of the bar was divided into a number of dark and dingy cubicles, each no longer than the straw mat on its floor, each with a crudely drawn number on its door. A single unshaded light bulb hung from the ceiling. I imagined the place by night, a small labyrinth of cells, each inhabited by a soft, pliant body.

A young boy appeared, and after a brief exchange with Rachel he led us out the back of the bar, past a number of small houses, and through a gate of rusted tin that marked the entrance of the hounfour. He knocked three times. Marcel Pierre emerged from his temple with a woman. She was short and languid. He was tall, with muscles that moved at bone level, and below his flat face, hidden by dark glasses, his thorax stood out in bold relief. He wore red, and a golf cap bearing the emblem of a manufacturer of insecticides.

As we had agreed on the drive, Rachel introduced me as a representative of powerful interests in New York who were willing to pay generously for his services provided no questions were asked and my instructions were followed precisely. He was not impressed. He fixed us with a long disconcerting stare, and when he began to speak he held his head stiffly as if conversing through an imaginary intermediary. He seemed uncommonly calm, and it soon became clear that his only concern was money—how much and how soon.

I asked to see a sample of the reputed zombi poison, and he led us into his *bagi*, the inner sanctum of the temple. An altar piled deep with artifacts took up most of the room. Displayed prominently among the brilliantly colored powders, the rum and wine bottles, the playing cards, feathers, heads, and Roman Catholic lithographs, were a doll's head and three skulls, one of them dog and two human. On the wall was a swollen carcass of a puffer fish, a sisal whip, and a staff decorated with horizontal bands of light and dark wood. Marcel reached into the pile on the altar and brought out a plastic bag containing a white aspirin bottle. From a ketchup bottle he poured an oily emulsion onto his hands and rubbed all exposed parts of his body, then instructed us to do the same. The potion smelled of ammonia and formaldehyde. Next, having wrapped a red cloth around his nose and mouth, he carefully opened the bottle. Inside was a coarse, light brown powder. Marcel stood back from the altar, and placidly lifted the red cloth from the left side of his face to reveal a mottled scar. This, he suggested, was proof of the powder's efficacy.

The negotiations began. He presented me with what amounted to a grocery list—so much for the zombi, so much for the poison, so much to dig up the necessary bones in the cemetery, so much for all three. His brusque offer, void of mystique, left me neither hopeful nor suspicious. By now, after but forty-eight hours in Haiti, I was becoming aware that in this surrealistic country anything might be possible. I merely insisted on certain conditions. I would pay him the negotiated price for the poison provided that I would be able to observe the entire process and collect raw samples of each ingredient. He hesitated, but then agreed. I told him that within twenty-four hours I would let him know if we needed any of his other services.

That evening when I was back with Rachel at the Peristyle de Mariani, I discussed Marcel's various propositions at some length with her father, Max. He assured me that an individual made into a zombi could be readily treated by a houngan. His confidence combined with my own skepticism that zombis even existed persuaded me to push Marcel Pierre as far as he would go. I would have a look at his poison, and if it was promising, other possibilities might follow.

We returned early the next morning, and Marcel Pierre took us to the local military post to request permission to enter the graveyard to dig up bones. This was denied, not for ethical reasons but because I had failed to obtain the necessary papers from the capital. Marcel suggested to me that we forget about obtaining fresh material from the cemetery and instead use some bones he had on hand at his temple. I agreed, and the three of us spent the rest of the morning assembling the various ingredients of the reputed poison. From an old apothecary we bought several packets of brightly colored talc—magical potions with such exotic names as "break wings," "cut water," "respect the crossroads." Then we drove north to a barren scrubland to gather leaves. We were back at the hounfour that afternoon, and beneath the thatch shelter of his peristyle Marcel Pierre prepared his zombi poison. He first ground the leaves in a mortar, then grated a human skull, adding the shavings to the mortar along with the miscellaneous packets of talc. It seemed a desultory process, his moving from task to task laboriously like an insect, drooping his shoulder with each step. It was near the end of the afternoon when he handed me a dark green powder, finely sifted and sealed in a glass jar.

Rachel cracked open the tin cap of a rum bottle, tipping it lightly to the ground three times to feed the loa. Marcel nodded approvingly. In between long swigs of rum I mentioned to Marcel that I planned to test the poison on an enemy I had, a white foreigner living in the capital, and that I would be certain to let him know the results. I thanked him profusely and paid him the substantial sum I had promised, plus a sizable bonus. I left his hounfour certain that he knew how to make the zombi poison. I was equally convinced that what he had made me was worthless.

When we drove up to the house at Mariani late that night, four men were waiting for us. Max Beauvoir introduced me but not them, saying simply that they wanted to know what I was doing in Haiti. Opening my pack, I laid Marcel Pierre's prepared poison on the table. One who seemed to be their leader, a short, gruff man with an enormous belly, took the poison, poured it into the palm of his hand, and stirred it with his index finger. Turning to Beauvoir, he said, "This is too light to be anything." Max laughed, and the others joined in. The four stayed long enough to have a drink and then left without waiting to see the ceremony.

"Who were they?" I asked Max as soon as they were gone.

"Important men."

"Houngan?"

He nodded.

I spent the next several days in the south looking for plantings of datura and the Calabar bean. All the species of datura that had been reported in Haiti were feral and quite weedy, and I had expected to find them growing in disturbed sites almost anywhere. Curiously, after walking the hills along the road over the mountains toward the southern port of Jacmel, and the barren fields along the east coast as far as Anse-à-Veau, I found but a single specimen—a scandent shrub of *Datura metel*, at a house site in a small coastal village, planted, I was told, as a remedy for asthma. As for the Calabar bean, I was equally disappointed. Having combed a number of low swampy habitats, and having perused the dusty herbarium at the Ministry of Agriculture, I found no evidence that the plant had become naturalized in Haiti.

But then, out with Max Beauvoir in the mountains above Port-au-Prince, I did find a species of the tree datura of the genus *Brugmansia*, planted as an ornamental. These are short, gnarly trees, almost invariably covered by large pendulous trumpet-shaped flowers. Though quite distinct in appearance from the spindly datura shrubs, they share the same active chemical principles and are equally toxic if ingested. This, in fact, was the same species used by the curanderos of northern Peru, the one known as cimora. I knew that the tree daturas were native to South America and had only recently been introduced into Haiti, but I was uncertain whether its special properties had been identified by the Haitian peasants. Apparently they had, for no sooner had I begun to collect specimens than a small rancorous group of hill peasants gathered, demanding to know why I was cutting that particular tree. But as soon as Beauvoir said a few words, their demeanor changed dramatically. They approached me expectantly, and a couple of young boys clambered up the trees for flowers. I turned to Beauvoir for an explanation.

"I told them that you are Grans Bwa, the spirit of the woods." He indicated an old woman, gnawing on a small pipe. "She has asked that you bathe them with herbs. I explained that we have no time. She argues that at least you must bathe the children. But come, I told her perhaps another time."

At that, I gathered my specimens and started down the hill toward the road. By then the ground at the base of the tree was blanketed with blossoms and the people had started to sing:

> *Leaves in the woods, call me*
> *Oh, leaves in the woods, call me*
> *Leaves in the woods, call me*
> *Ever since I was small, I have danced.*

Intrigued by the unexpected scarcity of datura in the fields of the south, I decided it was time to look up the last of my three leads on this increasingly enigmatic assignment. On a sultry afternoon when the capital smelled of spice, I contacted Lamarque Douyon, the psychiatrist who had been working with Clairvius Narcisse.

A secretary showed me into a stark office dominated by a massive mahogany desk and two photographs, each hung on a turquoise wall in frames that matched the desktop. One was a poster-sized presidential portrait of Dr. François Duvalier, a fixture that in most Haitian government offices has long since been replaced by a likeness of his son. The smaller photograph showed a young, almost unrecognizable Nathan Kline in a crewcut and horned-rimmed glasses. On another wall hung a plaque acknowledging the contributions of the American research institutions that had helped finance the initial construction of the clinic in 1959, and a second plaque thanking the foreign pharmaceutical companies for their contributions of free drugs for the first two years of operation. Beneath the barred windows stood a shock therapy unit, so archaic that it evoked a perverse nostalgia. The entire room was a freeze frame of the late fifties, the only time when research funds had flowed. Since then, not even the furnishings had changed.

Lamarque Douyon impressed me as a benign, soft-spoken man hampered not only by lack of funding but by the difficulties of reconciling his Western scientific training with the unique rhythms of his own culture and its thoroughly African foundations. He is a physician who straddles two very different worlds. As Haiti's leading psychiatrist and the director of its only psychiatric institute, he remains accountable to a peer group of foreign scientists; yet as a clinician he treads through an utterly non-Western landscape where European notions of mental health lose their relevance.

Douyon's scientific interest in the zombi phenomenon dates to a series of experiments he conducted in the late 1950s while completing his psychiatric residency at McGill University. That was the heyday of the early psychopharmacological research. Psychiatrists, having discovered that certain mental disorders could be successfully treated with drugs, were actively experimenting with a number of potent psychotropic substances on human subjects under controlled circumstances. What Douyon observed during some of these experiments reminded him of accounts of zombis he had heard as a child; he recalled as well the prevalent belief among many Haitians that zombis were created by a poison that brought on a semblance of death from which the victim would eventually recover. By the time he returned to Haiti in 1961 to take his position as director of the Centre de Psychiatrie et Neurologie, he, like Kline, was convinced that such a poison existed.

"Zombis cannot be the living dead," he told me. "Death is not merely the loss of bodily function, it is the material decay of the cells and tissues. One does not wake up the dead. However, those who have been drugged may revive."

"This zombi preparation, Dr. Douyon, do you have any idea what it contains?"

"Snakes, tarantulas, most anything that crawls." He hesitated. "Or leaps—they say there is always a large toad. Human bones . . . but they go into everything."

"Do you know the species of toad?"

"No, but I don't think . . . there is a plant that grows here. We call it the *concombre zombi,* the zombi's cucumber."

"Oh, you mean datura." I shared with him my thoughts about the Calabar bean.

He told me that he had never heard of the plant in Haiti and returned to the subject of datura. "When I was at McGill, my advisor suggested that I test a number of plants commonly associated with the zombi folklore. We fed a preparation of datura leaves to mice and induced a catatonic state for three hours. Unfortunately these experiments stopped when Professor Cameron left the institute."

"Was that Ewen Cameron?"

"Yes, you know of him?"

"By reputation." But I knew nothing of this connection. The late Ewen Cameron had been the director of the Allan Institution in Montreal during what Kline liked to refer to as the dark ages of psychiatry. Between 1957 and 1960 he conducted psychic driving and brainwashing experiments, funded partially by the CIA. He was notorious for combining LSD with massive doses of electroshock therapy in his studies of schizophrenia. After he left the institution, the subsequent director promptly banned all of Cameron's therapeutic techniques.

"Datura must be the principal ingredient in the poison," Douyon continued. "It is a powder that is placed in the form of a cross on the ground or across the threshold of a doorway. The peasants say that the victim succumbs just by walking over the cross."

"How much of this powder?"

"Very little, no more than a spoonful."

"Just on the bare ground?"

"The toxins must be absorbed through the feet," he concluded.

This I had trouble with. In East Africa certain tribes surreptitiously administer poisons by coating spiny fruits and placing them along footpaths. Datura is topically active, but without at least this type of mechanical aid it was difficult to imagine it or any substance passing through the callused feet of a peasant. It was also unclear how the poisoner, who was supposed simply to sprinkle the powder on the threshold of a hut, might ensure that only the intended victim suffered. I was curious whether Douyon had attempted to contact individuals responsible for making the zombi poison.

"I know a bokor in Saint Marc," he replied, "who supplied me with a vial of the poison, a white powder."

"Marcel Pierre?"

"You know him?" He seemed surprised.

"I got his name from the BBC. I was with him several days ago. Was that the white powder that you sent as a sample to Kline?"

"We had no facilities here to test it. Kline never sent me any results."

"There weren't any, at least nothing positive."

Douyon was disappointed, until I mentioned that the preparation obtained by the BBC had shown some biological activity.

"Did they get it from Marcel Pierre?" he asked. I nodded and he looked away. "Strange man, this Pierre. Did he tell you how many he has killed?"

"No."

Douyon frowned. "These people are criminals," he said. "It is always dangerous." He told me how he had once taken a foreign film crew to a cemetery in the town of Desdunnes to attempt to document a zombi being taken from the ground. They had been met at the entrance to the graveyard by an armed band of peasants who destroyed their cameras and placed them in jail overnight. He had never tried it again.

Douyon rose and started out to take a call. Before leaving he passed me a paper from a folder on his desk. It was a copy of a legal document, Article 249 of the Haitian penal code, that referred specifically to the zombi poison, prohibiting the use of any substance that induced a lethargic coma indistinguishable from death. It indicated that should a victim of such poisons be buried, the act would be considered murder no matter what the final result. The Haitian government apparently recognized the existence of the poison with some assurance. I had a look at several other papers in the folder. One was Clairvius Narcisse's medical dossier from the time of his death at the Schweitzer Hospital. I noted his symptoms: pulmonary edema leading to acute respiratory difficulties, rapid loss of weight, hypothermia, uremia, and hypertension.

I was not sure what to make of Douyon. On the one hand I had nothing but respect for his tenacious efforts, which after close to twenty frustrating years had finally yielded the provocative cases of Narcisse and Francina Illeus. Yet his pursuit of the reputed zombi poison had been less successful, and he seemed uncertain whether its formula would ever be discovered. He remained convinced that the active ingredient was datura, the same logical choice I had made when formulating my Calabar hypothesis. Yet despite his years of research, the only evidence that implicated datura was the fact that the plant was called the zombi's cucumber. The experiments he had conducted at McGill years before meant very little; any number of plant extracts injected into mice might induce a catatonic stupor. Datura might still be the prime candidate, but after more than twenty years Douyon had done little to test his hypothesis. He had yet to obtain a documented preparation of the poison.

There was another problem. As a Western psychiatrist he had redefined a zombi as a man or woman who—having been poisoned, buried alive, and resuscitated—manifested a certain set of symtoms. Yet the measure of a victim's psychiatric condition could never explain why an individual had become a victim in the first place. For example, in psychiatric terms, a zombi might be called a catatonic schizophrenic; both conditions are characterized by incoherence and catalepsy with alternate moments of stupor and activity. But catatonic schizophrenia as a syndrome exists worldwide and may result from a number of different processes that severely disrupt mental stability—including, perhaps, zombification. Certainly, if Narcisse's testimony was to be believed, we were talking about an extraordinary phenomenon that actually caused an individual to be buried alive. Such a traumatic experience might readily drive one mad. But a zombi represented far more than a set of symptoms; if true, zombification was a social process unique to a particular cultural reality. There had to be men and women actually creating this poison, deciding how, when, and to whom it should be administered, and completing the act by distributing and caring for the victims. Above all, if zombis actually existed, there had to be a reason, an explanation rooted in the structure and beliefs of the Haitian peasant society. Rather than seeking the cause and purpose of zombification within that traditional society, Douyon assumed zombification to be a random criminal activity and limited his efforts to observing and treating its effects. One point he had made about Francina Illeus's case had been particularly curious. When he attempted to return her to her home village, the family had refused to accept her. Douyon had explained that the people could not afford to feed an unproductive person. Yet in the short while that I had been in Haiti, I had already seen evidence that the elderly and infirm were well cared for by their families. I had an intuitive feeling that in her village, Francina represented far more than an extra mouth to feed.

A steel door opened behind me, and I heard the shuffling of bare feet on concrete. Douyon returned, trailed by a nurse and two patients. One of them I recognized.

When representatives of two completely different realities meet, words like *normal* become relative. I was in no position to judge if Clairvius Narcisse had been permanently affected by his ordeal. Physically he appeared fit. He spoke slowly but clearly. When questioned about his experience, he repeated basically the same account that I had heard from Nathan Kline. But he added certain extraordinary details. A scar he bore on his right cheek just to the edge of his mouth had been caused by a nail driven through his coffin. Quite incredibly, he recalled remaining conscious throughout his ordeal, and although completely immobilized, he had heard his sister weeping by his deathbed. He remembered his doctor pronouncing him dead. Both at and after his burial, his overall sensation was that of floating above the grave. This was his soul, he claimed, ready to travel on a journey that would be curtailed by the arrival of the bokor and his assistants. He could not remember how long he had been in the grave by the time they arrived. He suggested three days. They called his name and the ground opened. He heard drums, a pounding, a vibration, and then the bokor singing. He could barely see. They grabbed him, and began to beat him with a sisal whip. They tied him with rope and wrapped his body in black cloth. Bound and gagged, he was led away on foot by two men. For half the night they walked north until their party was met by another, which took custody of Narcisse. Traveling by night and hiding out by day, Narcisse was passed from one team to the next until he reached the sugar plantation that would be his home for two years.

Douyon lit a menthol cigarette for the second patient. She held it aimlessly, letting the ash grow until it dropped onto her lap. This was Francina Illeus, or "Ti Femme," as she was known. In April 1979 peasants from the Baptist mission at Passereine had noticed her wandering about the market at Ennery, had recognized her as a zombi, and reported her presence to the American in charge of the mission, Jay Ausherman. Ausherman traveled to Ennery and found an emaciated Francina squatting in the market with her hands crossed like kindling before her face. Three years before, she had been pronounced dead after a short illness. The judge at Ennery, uncertain of what to do with someone who was legally dead, willingly granted custody to Ausherman, who in turn passed her over to Douyon for psychiatric care. At that time she was malnourished, mute, and negativistic. For three years Douyon had attempted through hypnosis and narcosis to speed her recovery. He believed that there had been improvement. Still, her mental faculties were marginal. Her eyes remained blank, and every gesture was swollen with effort. She spoke now, but softly in a high thin voice and only when prodded gently by Douyon. There was little spontaneous emotion, and when she left the room she walked as if on the bottom of the sea, her body bearing the weight of all the oceans.

5

A Lesson in History

IN THAT FIRST WEEK in Haiti, and for several days that followed, I often spent mornings wandering restlessly between my room and the veranda of the hotel, picking up a pen only to drop it, a book only to leave it open on a table. I lied to tourists about who I was. After twelve days I still had nothing. Marcel Pierre's powder was clearly fraudulent. For now, there was little more to be learned at Douyon's clinic. The nation baffled me. Stunned by her multitudes, awed by her mysteries, dumbfounded by her contradictions, I paced. Only at dawn, when from sheer exhaustion or moved by the splendor of the city basking in such soft light, was I still. Sometimes with my eyes closed, and the silence broken only by the odd bird, I would hear whispered messages of the land that intuitively I understood, if only for a moment. Eventually I came to respect those moments, for the cycle of logical questions was getting me nowhere.

That was why I welcomed Max Beauvoir's suggestion that I forget about zombis and go with him to gather some leaves for his treatments. It is difficult and perhaps unimportant to capture the flavor of those outings, clouded as they now are by nostalgia. We took in the land, traveling its length, speaking constantly of its strengths, its weaknesses, its history, more often than not becoming lost in our thoughts and forgetting the purpose for which we had set out. Max Beauvoir placed the country before me like a gift. Images survive: streetside herbalists sheltered by ragged bits of awning, naked men in rice paddies, a string of peasants on a mountain trail, the angelic faces of their children, black as shadows. The days were fleeting and had a way of running into the night, and sometimes the singing never stopped and the drums called out painfully until one did not know which would be worse, to have them continue for another instant, or to have them stop. In the end, what emerged from these travels was a lesson in history, a lesson that served as a key to the symbols of the land.

In the closing decades of the eighteenth century the French colony of Saint Domingue was the envy of all Europe. A mere thirty-six thousand whites and an equal number of free mulattos dominated a slave force of almost half a million and generated two-thirds of France's overseas trade—a productivity that easily surpassed that of the newly formed United States and actually outranked the total annual output of all the Spanish Indies combined. In the one year 1789 the exports of cotton and indigo, coffee, cacao, tobacco, hides, and sugar filled the holds of over four thousand ships. In France no fewer than five million of the twenty-seven million citizens of the ancien régime depended economically on this trade. It was a staggering concentration of wealth, and it readily cast Saint Domingue as the jewel of the French empire and the most coveted colony of the age.

In 1791, two years after the French Revolution, the colony was shaken and then utterly destroyed by the only successful slave revolt in history. The war lasted twelve years, as the ex-slaves were called on to defeat the greatest powers of Europe. They faced first the remaining troops of the French monarchy, then a force of French republicans, before driving off first a Spanish and then a British invasion. In December of 1801, two years before the Louisiana Purchase, Napoleon at the height of his power dispatched the largest expedition ever to have sailed from France. Its mission was to take control of the Mississippi, hem in the expanding United States, and reestablish the French empire in what had become British North America. En route to Louisiana, it was ordered to pass by Saint Domingue and quell the slave revolt. The first wave of the invading force consisted of twenty thousand veteran troops under Bonaparte's ablest officers commanded by his own brother-in-law, Leclerc. So vast was the flotilla of support vessels that when it arrived in Haitian waters, the leaders of the revolt momentarily despaired, convinced that all of France had appeared to overwhelm them.

Leclerc never did reach Louisiana. Within a year he was dead, and of the thirty-four thousand troops eventually to land with him, a mere two thousand exhausted men remained in service. Following Leclerc's death, French command passed to the infamous Rochambeau, who immediately declared a war of extermination. Common prisoners were put to the torch; rebel generals were chained to rocks and allowed to starve. The wife and children of one prominent rebel were drowned before his eyes while French sailors nailed a pair of epaulettes into his naked shoulders. Fifteen hundred dogs were imported from Jamaica and taught to devour black prisoners in obscene public events housed in hastily built amphitheaters in Port-au-Prince. Yet despite this explicit policy of torture and murder, Rochambeau failed. A reinforcement of twenty thousand men simply added to the casualty figures. At the end of November 1803, the French, having lost over sixty thousand veteran troops, finally evacuated Saint Domingue.

That the revolutionary slaves of Saint Domingue defeated one of the finest armies of Europe is a historical fact that, though often overlooked, has never been denied. How they did it, however, has usually been misinterpreted. There are two common explanations. One invokes the scourge of yellow fever and implies that the white troops did not die at the hand of the blacks but from the wretched conditions of the tropical lands. Although without doubt many soldiers did succumb to fever, the supposition is contradicted by two facts. For one, European armies had been triumphant in many parts of the world plagued by endemic fevers and pestilence. Secondly, in Haiti the fevers arrived with the regularity of the seasons and did not begin until the onset of the rains in April. Yet the French forces led by Leclerc landed in February of 1802 and before the beginning of the season of fever had suffered ten thousand casualties.

The second explanation for the European defeat refers to fanatic and insensate hordes of blacks rising as a single body to overwhelm the more "rational" white troops. It is true that in the early days of the revolt the slaves fought with few resources and extraordinary courage.

Accounts of the time report that they went into battle armed only with knives and picks, sticks tipped in iron, and that they charged bayonets and cannon led by the passionate belief that the spirits would protect them, and their deaths, if realized, would lead them back to Guinée, the African homeland. Yet their fanaticism sprung not only from spiritual conviction but from a very human and fundamental awareness of their circumstances. In victory lay freedom, in capture awaited torture, in defeat stalked death. Moreover, after the initial spasm of revolt, the actual number of slaves who took part in the fighting was not that high. The largest of the rebel armies never contained more than eighteen thousand men. As in every revolutionary era, the struggle was carried by relatively few of those afflicted by the tyranny. The European forces suffered from fever, but they were defeated by men—not marauding hordes but relatively small, well-disciplined, and highly motivated rebel armies led by men of some military genius.

If the historians have clouded the character of the struggle, they have also inaccurately idealized the revolutionary leaders—Toussaint L'Ouverture, Christophe, and Dessalines, in particular—disguising their ambitions in lofty libertarian visions that they most certainly did not have. The primary interest of the French in the immediate wake of the uprising was the maintenance of an agrarian economy devoted to the production of export crops. How this was accomplished was of little concern. Once they realized that the restoration of slavery was not possible, and before Napoleon arose to attempt to storm the island by force, the French ministers devised an alternative system whereby freed slaves as sharecroppers would be forged into a new form of indentured labor. The plantations would essentially remain intact. Lacking the military presence to enforce this scheme, the French turned to the leaders of the revolutionary armies and found willing collaborators, including Toussaint L'Ouverture, who became a major figure in the restoration of French authority on the island. The French, however, made a critical error in assuming that this co-opted leadership would submit to the whims of Paris. On the contrary, the black leadership did what they had always planned to do. They secured for themselves positions at the top of a new social order.

Toussaint L'Ouverture had no intentions of overseeing the dismantling of the colonial plantations. In the abstract he was committed to the freedom of the people, but in practice he believed that the only way to maintain the country's prosperity and the free status of the citizens was through agricultural production. One of the most persistent myths about the Haitian revolution is the belief that the original plantations, having been destroyed in the initial uprising, never attained their former prosperity; the tacit assumption being that in the wake of the revolt the blacks who took over were incapable of governing. This is historically untrue. Within eighteen months of attaining power Toussaint L'Ouverture had restored agricultural production to two-thirds of what it had been at the height of the French colony. Had the French bourgeoisie been willing to share power with the revolutionary elite, it is possible that an export economy might have been maintained for some time. It was destined to collapse, however, not because of the lack of interest or inability of the new elite, nor even because of the chaos unleashed by Leclerc's invasion. Its eventual demise was assured by an expedient policy begun by the French long before the revolution in 1791.

The French plantation owners, faced with the difficulty of feeding close to half a million slaves, had granted provisional plots of land so that the slaves might produce their own food. The slaves were not only encouraged to cultivate their plots, they were allowed to sell their surplus, and as a result a vast internal marketing system developed even before the revolution. Thus the plantation owners in a calculated gesture had inadvertently sown the seeds of an agrarian peasantry. Yet another lingering myth concerning the revolution asserts that once the slaves were liberated from the plantations it was virtually impossible to entice them back onto the land. In fact, in the wake of the revolt, the majority of the ex-slaves went directly to the land, and energetically produced the staple foodstuffs that the internal market of the country demanded. Reading popular accounts of the twelve-year revolutionary war, one would assume that the entire population had scavenged for its sustenance. On the contrary, they were eating yams, beans, and plantains grown and sold by the majority of ex-slaves who cultivated their lands as freemen. The problem of the revolutionary elite was not to get the people back to the land, it was to get them from their own lands back to the plantations.

An independent peasantry was the last thing the black military leaders wanted. Jean-Jacques Dessalines maintained a dream of an export economy based on chain-gang labor up until his assassination in 1806. Henri Christophe, who ruled the northern half of the country until 1820, was temporarily successful, using measures every bit as harsh as those of the colonial era. For ten years he was able to produce export crops that allowed him to build an opulent palace and support a lavish court. Yet eventually his people revolted, and with his death in 1820, there disappeared the last serious attempt to create a plantation-based economy.

What emerged in the early years of independence was a country internally vigorous but externally quiescent. Productivity of export crops declined completely. At the height of the colony, over 163 million pounds of sugar were exported annually; by 1825, total exports measured two thousand pounds, and some sugar was actually imported from Cuba. Foreigners considered the economy to be dead, and again cited the inability of blacks to organize themselves. What these statistics in fact indicated was the unwillingness of a free peasantry to submit to an economic system that depended on their labor to produce export crops that would only profit a small number of the elite. The Haitian economy had not disappeared, it had simply changed. With negligible export earnings, the central government soon went broke. As early as 1820, then President Boyer was forced to pay his army with land grants. Thus unleashing the common soldier onto the land, he dealt the final blow to any lingering dreams of reestablishing the plantations. Recognizing that no income was going to accrue from non-existent export commodities, he began to tax the emerging structures of the peasant economy. In placing a tax on rural marketplaces, for example, he generated revenue, but more importantly from a historical perspective, he legitimized the institution itself. Then, unable to impose taxes or rents on the lands that the peasants had already taken as their own, he did the only thing that could raise income. He began officially to sell them the land. It was an extraordinary admission on the part of the central government that the peasants were in firm control of the countryside. The ex-slaves had moved onto the land, and nothing was going to pry them off it. The central government acquiesced and did what it could to generate at least some revenue from a situation that was totally beyond its control.

Yet who were these peasants who had so decisively rooted themselves to the land? Some perhaps were the descendants of the first slaves to arrive with the Spanish as early as 1510, but the majority had actually been born in Africa. Between 1775 and the outbreak of the revolution in 1791, the colony had expanded as never before. Production of cotton and coffee, for example, increased 50 percent in a mere six years, and to fuel such growth the slave population had been almost doubled. Yet because of the wretched conditions on the plantations at least seventeen thousand slaves died each year, while the birthrate remained an insignificant one percent. Hence, during the last fourteen years of undisturbed rule the French imported no fewer than 375,000 Africans. In other words, the germ of the modern Haitian peasantry quite literally sprouted in Africa.

The revolutionary slaves who settled the tortuous recesses of a mountainous island came from many parts of the ancient continent, and represented many distinct cultural traditions. Among them were artisans and musicians, herbalists, carvers, metalworkers, boatbuilders, farmers, drum makers, sorcerers, and warriors. There were men of royal blood, and others who had been born into slavery in Africa. In common was their experience with a heinous economic system that had ripped them away from their material world, but critically they also shared an oral tradition that was unassailable—a rich repository of religious belief, knowledge of music, dance, medicine, agriculture, and patterns of social organization that they carried with them into every remote valley. The evolution of these various traditions, their fusions and transformations, was deeply affected by a blanket of isolation that fell upon the country in the early years of the nineteenth century.

The nation that emerged from the revolutionary era was a pariah in the eyes of the international community. With the exception of Liberia, which was a limp creation of the United States, Haiti ranked as the only independent black republic for a hundred years. Its very existence was a constant thorn in the side of an imperialistic age. The Haitian government irritated the European powers by actively supporting revolutionary struggles that vowed to eliminate slavery. Simon Bolívar, for example, was both sheltered there and funded before he liberated Venezuela and the other Spanish colonies. In a more symbolic gesture, the government purchased shipments of slaves en route to the United States only to grant them freedom. Moreover, Haiti defied international commercial interests by prohibiting any foreigner from owning land or property within the country. By no means did this law bring trade to a standstill, but it dramatically modified its nature. In a century wherein European capital moved into virtually all regions of the world, Haiti remained relatively immune. Even the hegemony of the Roman Catholic church was checked. The clergy, which had never had a particularly strong presence in colonial days, lost virtually all influence after the revolution. In fact, for the first half-century of Haiti's independence, there was an official schism between the country and the Vatican. Roman Catholicism remained the official religion of the emerging political and economic elite, but during the seminal years of the nation, the church had practically no presence in the countryside.

Within Haiti, isolation of a different form occurred. Throughout the nineteenth century, as the colonial infrastructure of roads decayed and was not replaced, the physical gap between town and country widened. This, in turn, sharpened an emerging cultural hiatus between two radically different segments of Haitian society, the rural peasants and the urban elite. The former, of course, were ex-slaves; the latter, in part, descendants of a special class of free mulattos that during the colonial era had enjoyed both great wealth and all the rights of French citizens, including the ignoble privilege of owning slaves. During the early years of the independence, the obvious differences between these two groups crystallized into a profound separation that went far deeper than mere class lines. They became more like two different worlds, coexisting within a single country.

The urban elite, though proudly Haitian, turned to Europe for cultural and spiritual inspiration. They spoke French, professed faith in the Roman Catholic church, and were well educated. Their women wore the latest Parisian fashions, and their men naturally formed the chromatic screen through which European and American commercial interests siphoned what they could of the nation's wealth. Their young men frequently traveled abroad, for both higher training and amusement, invariably returning to fill all business and professional positions, as well as governmental and military offices. There they promulgated the official laws of the land, all of which were again based on French precedent and the Napoleonic Code. By all foreign standards, it was this small circle of friends and extended families—for the elite never numbered more than 5 percent of the population—that controlled much of the political and economic power of the nation.

In the hinterland, however, the ex-slaves created an utterly different society based not on European models, but on their own ancestral traditions. It was not, strictly speaking, an African society. Inevitably European influences were felt, and only very rarely did pure strains of specific African cultures survive or dominate. What evolved, rather, was a uniquely Haitian amalgam forged predominantly from African traits culled from many parts of that continent. Typically, its members thought of themselves less as descendants of particular tribes or kingdoms than as "ti guinin"—Children of Guinée, of Africa, the ancient homeland, a place that slowly drifted from history into the realm of myth. And, in time, what had been the collective memory of an entire disenfranchised people become the ethos of new generations, and the foundation of a distinct and persistent culture.

Today evidence of the African heritage is everywhere in rural Haiti. In the fields, long lines of men wield hoes to the rhythm of small drums, and just beyond them sit steaming pots of millet and yams ready for the harvest feast. In a roadside settlement, or *lakou*, near the center of the compound, a wizened old man holds court. Markets sprout up at every crossroads, and like magnets they pull the women out of the hills; one sees their narrow traffic on the trails, the billowy walk of girls beneath baskets of rice, the silhouette of a stubborn matron dragging a half-dozen donkeys laden with eggplant. There are sounds as well. The echo of distant songs, the din of the market, and the cadence of the language itself—Creole—each word truncated to fit the meter of West African speech. Each of these disparate images, of course, translates into a theme: the value of collective labor, communal land holdings, the authority of the patriarch, the dominant role of women in the market economy. And these themes, in turn, are clues to a complex social world.

Yet images alone cannot begin to express the cohesion of the peasant society; this, like a psychic education, must come in symbols, in invisible tones sensed and felt as much as observed. For in this country of survivors and spirits, the living and the dead, it is religion that provides the essential bond. Vodoun is not an isolated cult; it is a complex mystical worldview, a system of beliefs concerning the relationship between man, nature, and the supernatural forces of the universe. It fuses the unknown to the known, creates order out of chaos, renders the mysterious intelligible. Vodoun cannot be abstracted from the day-to-day lives of the believers. In Haiti, as in Africa, there is no separation between the sacred and the secular, between the holy and the profane, between the material and the spiritual. Every dance, every song, every action is but a particle of the whole, each gesture a prayer for the survival of the entire community.

The pillar of this community is the houngan. Unlike the Roman Catholic priest, the houngan does not control access to the spirit realm.

Vodoun is a quintessentially democratic faith. Each believer not only has direct contact with the spirits, he actually receives them into his body. As the Haitians say, the Catholic goes to church to speak about God, the vodounist dances in the hounfour to become God. Nevertheless, the houngan's role is vital. As a theologian he is called upon to interpret a complex body of belief, reading the power in leaves and the meaning in stones. Yet vodoun not only embodies a set of spiritual concepts, it prescribes a way of life, a philosophy and a code of ethics that regulate social behavior. As surely as one refers to Christian or Buddhist society, one may speak of vodoun society, and within that world one finds completeness: art and music, education based on the oral transmission of songs and folklore, a complex system of medicine, and a system of justice based on indigenous principles of conduct and morality. The houngan as de facto leader of this society is at once psychologist, physician, diviner, musician, and spiritual healer. As a moral and religious leader, it is he who must skillfully balance the forces of the universe and guide the play of the winds.

Within the vodoun society, there are no accidents. It is a closed system of belief in which no event has a life of its own. It was within this society that Clairvius Narcisse and Ti Femme became zombis.

"Look into the sky and what do you see?" Rachel asked, staring far into the darkness. There was a small cooking fire between us, and in the flamelit smoke her face softened, her skin flushed in copper.

"Stars, sometimes."

"When I was small my father took me to a planetarium in New York. You have millions of stars, and your astronomers have even more." She stood up and walked slowly into the shadow, her words falling away like sparks into the night.

"Look into this sky. We have only a few, and when the clouds come in even fewer. But behind our stars, we see God. Behind yours, you just see more stars."

Her words saddened me unexpectedly, exposing as they did the gap that lay between me and her people. I gazed down the slope and across the crowded valley alive with twinkling fires, and followed the movements of a torch beam as it climbed erratically toward the crest of a draw. I remembered a recent day when her father and I had been out walking. We had come to a height of land where we could look across a scorched valley stripped to the bone, with haze rising off hot white stones and a few gnarly native trees among the ubiquitous thornscrub and neem. Max Beauvoir had taken a deep breath as if the very sight of such a landscape might bring tears to his eyes—and it probably could have. He had waxed eloquent as if words alone might have squeezed beauty out of that wretched sight, from the wasteland created by years of neglect. It had been extraordinary. I could only think of locusts, he of angels, yet who was the wiser?

Like all contented men, Max Beauvoir had by middle age found his rest. It had not come easily. The son of a bourgeois physician, he had left Haiti as a youth on a remarkable odyssey that led him from the streets of New York to the Sorbonne in Paris and finally to the court of the king of Dahomey. After fifteen years abroad he returned to Haiti, a chemical engineer intent on growing sisal, from which he would extract cortisone. His plantings were barely established when his work was interrupted by the death of a grandfather, a houngan, who from his deathbed instructed Max to take on the mantle of the religious tradition. Soon after, his commercial venture abandoned, Max Beauvoir began a second journey. For five years he crisscrossed Haiti observing the traditional rites, partaking in the pilgrimages and listening to the houngans. Finally, for a spiritual parent he chose an old man from the south and joined the community of his hounfour, eventually becoming initiated as a houngan himself. Into that world he led his wife, a lovely French painter, and perhaps more reluctantly their two daughters. The young girls were teased by their classmates because of their father's beliefs. At first they had been ashamed, but as they grew older it became their greatest source of pride.

"My father's great-grandfather had green eyes," Rachel said, returning to the fire, "and a gold watch. He came from the east, on horseback and with everything he owned stuffed into a calabash that hung around his neck. He could take that watch—it had a mirror on the inside—and just by opening it and staring into the mirror, he would disappear." She glanced at me, as if uncertain whether I would believe the story.

"He wandered all over the country and finally settled in Petite Rivière de l'Artibonite. He became really important."

"Aided no doubt by his watch," I teased.

"Perhaps." She smiled. "He was a great houngan. He lived for many years, but he knew when he was going to die. Just before, he gave away everything he owned and called together all the family.

They found him beneath a great *mapou* tree, and one by one he called for his grandchildren, including Max's father. He said something to each one and then he just left."

"Where to?"

"No one knows. He placed a few things in a saddlebag, and rode away. They say towards the Dominican Republic."

"That was the end?"

"Yes. Can you imagine? No one ever knew his birthplace or where he died."

I looked across the fire at her but said nothing. For weeks the two of us had been as blind accomplices inadvertently placed on the same trail pursuing different goals. I was chasing an elusive phenomenon that I scarcely believed in; she sought a sense of place. For she was young, and whether she knew it or not she was torn between two worlds, that of her lineage, which she bore like a weight, and everything that awaited out there beyond the ocean. One knew just by looking at her that she would have to go sometime. She had a hundred precocious ideas, and some were good and true, but they could never be hers until she found them alone, for ideas are but words unless they are sown in experience. At the same time I was drawn to warn her somehow, for hers was a beauty that filled one with a premonition that it could easily become the target of destructive forces. Yet you could not feel sorry for her; her pride forbade it. And moreover, nothing evil had yet befallen her; she lived with the brash confidence of a person who has never lost.

The dusty road to Savanne Carée passes through the old market center of Ennery, with its cobbles and thick-walled caserne and forlorn statue of Toussaint L'Ouverture, then runs along the Rivière Sorcier, crisscrossing a number of affluents before climbing into a high, rolling landscape dappled with mango trees. It is rich land with clusters of white houses, orchards swollen in fruit, and fallow slopes camouflaged by thick grass. Millet grows to the edge of the valley where the fields become broken, and the skirt of the mountain rises to steep cliffs draped in cascading remnants of native vegetation. In the hedgerows there are even wild things, and overhead the odd raptor scrapes the sky. After the barren lands to the west around Gonaives, there is an innocence to such abundance. Yet somewhere in this picturesque valley Ti Femme squandered her birthright and was made a zombi.

Just beyond a crossroads littered with the remnants of the morning market, Rachel and I noticed a young lad prostrate on the stony ground, relaxed as a lizard. He must have seen us at the same time. He leapt to his feet and in an instant was joined by another. As we came nearer, I saw that the first was a small wisp of a boy, with electric eyes and the large head and long limbs typical of many peasant youths. This was Oris. The other was René, and he was younger and clearly subordinate. From that moment on we were four.

"Of course, I know her. The one that passed beneath the ground," Oris said to Rachel as we made our way along the narrow trail. "She's the one who died, but her breath of life didn't leave, it stayed here," he added, pointing to his armpit.

"They buried her?"

"Naturally, and then at night the person who killed her went to get her." René was running up and down the trail like a rabbit anxious not to miss anything. Every so often his excitement overflowed in laughter, gay and sparkling.

"Who did this to her?"

"We don't know," René said, glancing ahead toward his friend.

"Her aunt," Oris said matter-of-factly. "Then they put her to work cleaning cornmeal. That's all she was good for."

"And the bokor?"

"You can't know. To find out you'd have to know who killed her."

"I thought you said the aunt."

"She had many aunts, and some have died. Besides, nobody cared anyway."

"What do you mean by that?" Rachel paused, squatting by the side of the trail, her skirt spread out from her waist like a tent. Oris swaggered by her unabashed. A market woman with a healthy corpulence and a lugubrious air was walking up the trail toward us, and Oris didn't say anything until she had passed.

"Ti Femme was mean. She was doing everything she shouldn't have done. And she didn't like people."

"Why? What did she do?" Rachel asked.

"Well, for example, you don't do anything and she stands up and swears at you. She swore at people for no reason. But here they kill people for almost nothing," Oris said casually. "It's just like the woman in the elections," he continued, referring to the recent political campaign for the National Assembly. "We didn't like her. She was ill behaved and had meetings and swore at the children. So we got her. I, myself, voted five times for the doctor. He promised to bring the president's wife to our lakou."

We followed our insouciant guides into a dense canebrake and then made our way slowly until reaching a narrow path that pointed to a small house perched on a high mound of earth. In the shade of its porch two men were pounding grain in a mortar, the rhythmic movement of the pestles fluid and powerful. Oris nodded to Rachel, then he and his companion backed away and the reeds closed behind them.

"*Honneur!*" Rachel called out as we approached. The men glanced up, and one of them sent a young child scurrying into the house. They continued to look at us for a while, then returned to their work. It was a crude welcome, far below the standards of peasant etiquette. We waited beneath a tree laden with green oranges until a frail old woman appeared. She was the mother of Ti Femme, and her name was Mercilia. We gave her ours, and she invited us into the front chamber of her home, where we shared cigarettes and coffee. After several minutes of perfunctory chatter, we came to the point of our visit.

"The girl died by God's will," Mercilia said, "and was revived by God's will. We know nothing of these things. They are wrong to say someone killed her. She died, she just died."

"Yes, that is what we have heard. And she had been quite ill, I suppose."

"For months, fever and a pressure from beneath her heart. She was a good girl, everyone liked her. They all came to the funeral. There were two wakes." She nodded confidentially toward Rachel. "The coffin didn't come, and then it rained."

Rachel cast a perplexed glance my way, then turned quickly back to Mercilia. "Where was she when she died?"

"Right here in the house of her birth. In the night. They buried her at daybreak. I went to the grave three days after." She was rocking slowly on the edge of her chair. "Ti Femme never said anything about passing through the earth, only that she heard people say she was dead like she was dreaming and she couldn't do anything. When I went I didn't know she wasn't there."

"Where is she now?" Rachel asked.

"With the state, in the clinic. If I could, I would have her here, but my husband is dead and Ti Femme has no one and is a child. When they found her, she couldn't even bathe or comb her own hair."

The door creaked open revealing a pod of children eavesdropping. Mercilia shooed them away. A pair of chickens slipped past her legs and sped into the room, clucking and pecking the cement floor out of habit. Rachel and I stood up to leave.

"There was not a single person who didn't like her," Mercilia volunteered, as she took a little money from my hand. "She never had arguments with anyone."

As expected, our talkative guides were waiting for us just beyond the canebrake. Oris was stretched out in the shade of a *mombin batard* tree, resting on one elbow and gnawing on a stick of sugarcane.

"Well?" he asked with puckish confidence.

"Well what?" Rachel replied, laughing.

The walk back to our jeep was uneventful, and once we had dispatched the two lads at the marketplace Rachel turned to me.

"It's odd, you know. A wake is always held at night, and if there were two the body must have been waiting around for what, thirty-six hours?"

"At least."

"Wouldn't the family have noticed that it hadn't begun to decompose?"

"You'd assume so." I thought for a minute. "What do you make of the boy's story."

"I'm not sure. But there is a friend of my father's on the coast who trades here, and she would know."

The sun was just going down when we reached the outskirts of Gonaives and pulled up to a dusty compound enclosed by a tall blue-and-green wall decorated with a great naive painting of a mermaid.

"It's a nightclub sometimes," Rachel explained. "Clermezine is the mermaid, it's one of her spirits. She's a great *serviteur*." Rachel said something to one of the idlers hanging around the entrance. He slipped inside and returned in a moment followed by a young woman. Rachel kissed her gently on both cheeks, and we followed her past a concrete dance floor and a broken-down bandstand to the inner courtyard. To one side some kind of noisy cabal was under way, presided over by a most extraordinary woman. When she stood up to receive us she was as regal and imposing as a queen. Rachel disappeared beneath a heap of endcarments, and before we had a chance to find the places made for us in the small circle of chairs, a tray of steaming thick coffee arrived.

Within a few moments a great whirlwind of voices enveloped us. I had no idea what was going on. The woman in charge seemed in deadly earnest, and one of the men grew steadily angrier, but Rachel was beside herself with laughter. From what I later gathered the man had been entrusted with some task, which he had failed to carry out. The woman suggested that if he let her down again, "there won't be a shovel small enough to pick up your pieces." He in turn threatened to send a *loup garou,* a werewolf, her way. She countered by saying that she could fly faster than anything he could come up with, especially if, like him, it was hindered "by those things that hang between your legs." This nattering salacious humor continued for some time, until finally the woman bellowed a few harsh words that cut him off immediately.

The subject of Ti Femme set her off again. She explained that Ti Femme used to come down to Gonaives to buy cornmeal, which she then sold at a profit in Ennery and Savanne Carée. She called her *maloktcho,* a Creole invective that translates poorly as "crude, uncivilized, raw." Like young Oris, she said that Ti Femme was rude and always swore at people. She was also dishonest.

"If you went to buy from her, and what she was selling was worth five gourdes, she'd say seven, then six, and you'd say five and she'd take it. But when she had measured it out, she'd hand it to you and say six. That's why they killed her."

"Some say her family was behind it."

"I tell you it was in the market. Everyone hated her. If you left your money out, she'd take it. She was a thief."

"It could have been anyone, then?"

"All of them! No one person could afford to kill her."

The next morning we passed by the Baptist mission at Passereine, intending to speak with Jay Ausherman, the American woman who had cared for Ti Femme just after she was found in the Ennery marketplace in 1979. The missionary was out of the country, but as we drove away I noticed a robust, balding man sitting alone on the steps of the cinderblock church. His was not a face readily forgotten. As we soon discovered, Clairvius Narcisse had been living off and on at the mission since being discharged from the psychiatric institute.

In this, the first of a number of informal interviews that took place away from Douyon's clinic, Narcisse spoke more easily about his ordeal. He had been a very strong man, and almost never sick, he claimed, and he hadn't suspected anything. There had been a dispute with one of his brothers, a bokor who coveted a piece of land that Narcisse had been cultivating, and only now did he fully understand what had occurred. His brother passed the magic to him on a Sunday. Tuesday he had been in Gonaives, feeling weak and nauseated. By the time he entered the hospital late that day, he was coughing and having difficulty breathing. By noon the next day, he was dying.

"What was this poison they passed onto you?" Rachel interrupted.

"There was no poison," he replied, "otherwise my bones would have rotted under the earth. The bokor sent for my soul. That's how it was done."

"In the basin?" she asked.

"Yes. It's full of water, but they prick your skin and call the spirit and the water changes into blood."

Narcisse explained that he had been sold to a bokor named Josef Jean who held him captive at a plantation near Ravine-Trompette, a small village in the north close to Pilate, not far from Cap Haitien. Together with many other zombis, he had toiled as a field hand from sunrise to sunset, pausing only for the one meal they received each day. The food was normal peasant fare, with the one restriction that salt was strictly prohibited. He remembered being aware of his predicament, of missing his family and friends and his land, of wanting to return. But his life had the quality of a strange dream, with events, objects, and perceptions interacting in slow motion, and with everything completely out of his control. In fact, there was no control at all. Decision had no meaning, and conscious action was an impossibility.

His freedom had come about quite by chance. One of the captives had refused to eat for several days and was beaten repeatedly for insubordination. In the midst of one such beating, the zombi got hold of a hoe and in a fit of rage killed the bokor. With the death of the master, the zombis dispersed, some eventually returning to their villages scattered across the northern plain. Only two of them were from the middle of the country—Narcisse and one other, who curiously enough came from Ennery. After being set free, Narcisse remained in the north for several years, until moving south to Saint Michel de l'Attalaye, where he settled for eight years. Although fear of his brother kept him from returning to his village, he did attempt to establish contact with his family. His many letters, however, went unanswered. Finally, when he heard that the brother responsible for his ordeal had died, he returned to l'Estère. His arrival, not surprisingly, had shocked the community. In truth, he confessed, he had not been well received. The villagers had taunted him, and such was the commotion that the government authorities deemed his life in danger and placed him in jail for his own protection. It was at that point that he had come under the care of Dr. Lamarque Douyon. To this day, he returns to l'Estère only for brief visits. His time is spent either at Douyon's private clinic in the capital or in the refuge of the Baptist mission.

"They called my name three times," Narcisse told us later that afternoon as he sat in the cemetery at Benetier. Upon our approaching the cemetery, it had taken Narcisse several minutes by the side of the road to become oriented. Then with little hesitation he had woven his way through the crowded tombs of cracked concrete until he reached his own. Scarcely visible, etched into the surface of the cement, was an epitaph written some twenty years before: "Ici Repose Clairvius Narcisse."

"Even as they cast the dirt on my coffin, I was not there. My flesh was there," he said, pointing to the ground, "but I floated here, moving wherever. I could hear everything that happened. Then they came. They had my soul, they called me, casting it into the ground." Narcisse looked up from the ground. At the edge of the cemetery a pair of thin gravediggers stood as still and attentive as gazelles. Narcisse felt the weight of their recognition.

"Are they afraid of you?" Rachel asked.

"No," he replied, "only if I was creating problems, then I'd have problems myself." He didn't say anything more for several minutes. The late afternoon light illuminated his face but left a conspicuous dark spot over the deep scar in his right cheek. It was there, he had mentioned earlier, that the nail of the coffin had pierced his flesh.

"They thought I was a *bourreau* [an executioner], so after they passed the bottle, they bound my arms to my sides."

"Did you have the force to resist?"

As if he hadn't heard me, Narcisse went on, "Then I was taken for eight days of judgment."

"By whom?" Rachel asked excitedly, "where did it take place?"

Again Narcisse ignored the question, and began to make his way out of the graveyard. He paused momentarily by a large erect tomb, and then continued to the road. As we reached the jeep, he turned to us both and said very quietly. "They are the masters of the country, and they do as they please."

"The only tribunal that my brother knew was the cemetery." Angelina Narcisse sat back in her chair, her legs wide apart. The morning sun had conquered the clouds and driven us into the shadows of the thatch shelter. Between the thin rafters ran long strings displaying dozens of photographs of President Jean-Claude Duvalier and his wife. Michelle Duvalier's face stood out, polished and quite lovely. Between the photographs were small Haitian flags, red and black, the same colors as Angelina's long dress. All around us the houses of the lakou stood as one, fused like stones encased in dry clay. In one corner iron rods had been set into the earth, at their base a pile of black coals. Once while out with Beauvoir I saw a possessed man implant such a rod in the ground, his bare hands wrapped around the iron, its tip red-hot, his face indifferent.

"Unless, of course, it was a tribunal at night, in which case nobody should know of it. So we don't know of it."

In a harsh voice, Angelina laid out her version of her brother's case. Long before the death of their parents, Clairvius had been involved in innumerable disputes with his various brothers. Land was often an issue, but there were others. Clairvius had done well financially, but showed no willingness to spread his earnings through his family. Once, for example, his brother Magrim sought a twenty-dollar loan, which Clairvius refused categorically. An intense argument had followed, which had culminated in Magrim striking Clairvius in the leg with a log, while Clairvius responded by hurling stones. Both of them had ended up in jail.

Apparently Clairvius had antagonized not only his family. He had compromised innumerable women, scattering children to all corners of the Artibonite Valley. None of these he accepted responsibility for, nor had he built houses for the various mothers. As a result he had approached middle age with few financial burdens, which freed him to advance further than his more responsible peers. He placed a tin roof on his house, for example, before anyone else in the lakou. Clairvius had profited at the expense of the community, and in all likelihood, suggested Angelina, it was one of the aggrieved members, probably a mistress, that had sold him to a bokor.

"But we know nothing of poisons in our family," she concluded. "My brother was sick for a year. It was not a disease from God, and there was no poison, or he would still be in the ground."

Whatever the cause of her brother's demise, the family lost no time taking over his fields, which Angelina and another sister still work today. Although Clairvius has made a claim in the national courts for his land, his sisters have absolutely no intention of releasing anything to him. As far as they are concerned Clairvius remains a dead man, a spirit that should never have returned to the village. In fact, the first member of the family to recognize him when he appeared in the l'Estère market in 1980 had sent for Angelina, and then told Clairvius to go away. Another sister arrived from the lakou and offered Clairvius money, but also ordered him to leave the village. By then a great crowd had gathered, and the police arrived to take Clairvius Narcisse to the protection of the government jail.

Death in a family should be like a stone cast into a lake; it makes a brief hole, but the waves of sorrow reach to the edge of the bloodline. In the case of Clairvius Narcisse, however, the stone slipped into the water without leaving a trace. Not long after leaving the family lakou, we discovered why. In the searing midday sun, we pulled off the road to offer a ride to a solitary peasant burdened by a ponderous load. Quite by chance he was a cousin of Clairvius, and with little difficulty Rachel persuaded him that we knew more than we did.

"Of course it was someone in the family, that is certain," the man told us. "But you'll never know what he did unless you can speak with the one who judged him."

"But the houngan said that the tribunal was never summoned," Rachel replied, leading him on deliberately.

"There must be a tribunal in a case like that," the cousin insisted. "They must call the dead. Otherwise they cannot set the trap."

"To take his soul?" I asked, remembering what Clairvius had told us at the mission.

"They must."

"His sister said the bokor passed a *coup l'aire*," Rachel said.

"No. They wanted the body for work. Only a *coup poudre* could bring him down."

"Coup poudre?" I looked toward Rachel.

"A magical powder," she explained.

The cousin was uncertain what was in it. A large lizard called the *agamont*, perhaps, and apparently two toads, the *crapaud bouga*, and another called the *crapaud de mer*, the sea toad.

"Where is the poison placed?"

"There is no poison," he answered. "Narcisse came out of the ground. A poison would have left him where he lay." The cousin looked perplexed, and suddenly I understood and felt terribly foolish. I had been asking for a poison, but what I called a poison they called a trap or a coup poudre. For them what created a zombi was not a drug but a magical act.

It was two in the morning before the other guests left the Peristyle de Mariani, and Rachel went inside, leaving her father and me alone on the terrace. In the wake of the ceremony and the ebullient conversation that invariably followed, there was a welcome stillness that allowed one to notice things, like the fragrance of the lemon trees, or the high whistle of the fruit bats.

"You are not in your plate tonight," Max said, drawing me back into his own realm of words. A bat swept beneath the roof of the peristyle and in an instant was gone.

"Perhaps," I said offhandedly.

"Do you want a drink?"

"If you do. Rum on ice."

"How's the work?"

"It goes. Rachel didn't tell you?"

"Some of it. They seem to be quite a pair."

"Ti Femme?"

"And the other one."

"Max, somebody has it all wrong."

"What do you mean?"

"Ti Femme wasn't innocent. People hated her. And Narcisse's own sister tried to throw him out of the village, after not having seen him for eighteen years."

"Probably she was afraid of him."

"How can you be afraid of something that has no will?"

Beauvoir had started to laugh, but what I said stopped him. His face turned like a mask into a different pose.

"And how," I continued, "can a being without will deliberately choose to kill someone? That, according to Narcisse, is how they gained their freedom. It doesn't make sense."

"You don't believe him?"

"I don't believe he was an innocent victim. Neither of them. Narcisse said there was a judgment, so did his cousin."

Beauvoir didn't say anything for several minutes, and when he did there was an unfamiliar edge to his voice. "You spoke to his sisters?"

"Just the tall one, Angelina. We saw the others some time ago."

"What did you make of her?"

"Powerful, totally in charge. No love lost for her brother."

"Nothing peculiar? Think about it."

"No." I paused for some time. "Except, I suppose, when I asked to photograph the family."

"And?"

"She was the only one to change her dress."

"So you did notice."

"Yes, but I don't . . ."

"She is a queen, and you surprised her." I began to interrupt again, but he didn't let me. He lifted his drink and the light caught the edge of the glass.

"To understand Haiti," he said, "you must think of a glass of water. You cannot avoid touching the glass, but it is just a means of support. It is the water that slakes your thirst and it is the water, not the glass, that keeps you alive.

"In Haiti the glass consists of the Roman Catholic church, the government, the National Police and army, the French language, and a set of laws invented in Paris. Yet when you think of it, over ninety percent of the people do not understand, let alone read, French. Roman Catholicism may be the official religion, but as we say the nation is eighty-five percent Catholic and one hundred and ten percent vodoun. Supposedly we have Western medicine, but in a country of over six million, there are but five hundred physicians, and only a handful of these practice outside of the capital.

"No, from the outside Haiti may appear to be like any other forlorn child of the Third World struggling hopelessly to become a modern Western nation. But as you have seen, this is just a veneer. In the belly of the nation there is something else going on. Clairvius Narcisse was not made a zombi by some random, criminal act. He told you he was judged. He spoke about the masters of the land. Here he did not lie. They exist, and these are the ones you must seek, for your answers will only be found in the councils of the secret society."

6

Everything Is Poison,
Nothing Is Poison

I COULD SEE THE horizon stained by the sea, and the first shafts of light to the east. The air was cool and serene. It was a special time when the city shifted its mood, when the people emerged as the night angels fled, and the light made the buildings blush.

I still didn't know his name, nor did I want to. He had a way of appearing at the strangest moments, always in linen, the same black cane in his hand, a jester's baton, tapping the alabaster steps of the hotel.

"Ah, my young friend, you do not sleep."

"Not tonight. I am just in."

"And how fares your hunt?" I had told him nothing of my work. "I see by your expression. A shame. Mine, too, I'm afraid. A sordid commerce, trading one's dollars for dreams. You could do better among the women of the Carrefour Road."

His thin fingers plucked a tired rose from his lapel and discarded it on the table beside me. Once I had seen him in the bar of the hotel passing out copies of an article about himself in an American midwestern newspaper. It was ironic to see someone who exuded such self-importance striving so obviously for notoriety. Like the entire nation he was hungry for recognition. Yet it was he who, one afternoon with the heat rising off the veranda, had told me, "The world is not after Haiti as so many of us feel. The cold truth is the world's indifference, and if there is one thing a Haitian hates it is to be inconsequential. It does not matter what is said about you, as long as you are the subject of conversation. Perhaps at some international soiree idle chatter passes to Haiti, but I doubt it." One wanted to avoid him, yet in a country where the truth is taken so lightly, casually, he was irresistible.

When he was gone I moved to the edge of the veranda and looked out over the city. It was hard to believe that in a few hours the heat and glare of the sun would be difficult to bear.

Marcel Pierre was the only person I knew capable of making the zombi poison. In the three weeks since we had left his hounfour I had not, as promised, tested his preparation on an enemy. I had, however, learned something about the man. According to a number of informants Marcel Pierre was an early and loyal follower of François Duvalier, and a notorious member of the Ton Ton Macoute, the rural militia that Duvalier established to safeguard his regime. Apparently during the terrible days of the early 1960s, when to consolidate his revolution Duvalier killed hundreds of the mulatto elite, Marcel had used his authority to extort information from the traditional houngan. Although many of the houngan were themselves members of the Ton Ton Macoute, few could match Marcel Pierre's ruthlessness. In exchange for protection from his gang, they reluctantly took him into their fold and shared certain secrets. Eventually Marcel's abuses became intolerable, and he himself fell victim and was severely poisoned. Though he barely survived, the poison left him permanently scarred. Now close to twenty years later, his reputation was mixed. Some said he had repented and become a legitimately initiated houngan; others described him as a houngan *macoute*, a charlatan who lacks true spiritual knowledge. Still others, perhaps the majority, dismissed him as a lowly *malfacteur*, an evildoer, just as the BBC had done.

Though Marcel and his followers had a reputation for selling the powders, it had been immediately apparent to me that the preparation he had made me was fraudulent. The plants used belonged to botanical families known to have little phytochemical significance. And then there was his attitude. He had prepared the reputed poison in the presence of young children and in the immediate vicinity of the living quarters of the houngan. This was most suspicious. Biodynamic plant preparations, be they poisons or hallucinogens, are inherently subversive—the poisons because they kill surreptitiously and the hallucinogens because they expose the frail, ambivalent position of man, perpetually on the cusp between nature, society, and the spirit world. When the shamans in the Amazon prepare and imbibe their potent hallucinogens, they usually leave the village and take their patients with them. Similarly the curare, or arrow poison preparations, are often made in the forest. Haiti is not the Amazon, but I was certain that Marcel Pierre, whatever his reputation, would not make such a deadly, topically active poison so close to his temple and the homes of his people. That would make no sense. Yet if Marcel knew the correct formula, how were we to persuade him to give it to us?

I did have one advantage. Other investigators had assumed that the formula of the zombi poison had to be an esoteric secret. I wasn't so sure, and anyway I had come not to believe in secrets. For years in the Amazon I had sought information that though not explicitly termed secret was nevertheless closely guarded by Indian peoples. In my experience, success depended less on the inherent status of the information than on the relationship one managed to establish with a knowledgeable informant. Every society has codes and rules, but individuals within a society bend them with zeal. Secrecy may be a rule, but gossip, after all, is the international language.

The purpose of secrets is to protect a society's interests from threatening outsiders. If a relationship can be established that renders one no longer a threat, the need for the protective veil vanishes. Sometimes such a relationship can take months to establish. Once I spent two seasons attempting to record some tribal myths from an old Tsimshian Indian in northern Canada. I had completely given up, concluding, as he had implied, that the tradition had been lost. Then one day, simply as a gesture, I shot and butchered a moose for him for his winter meat. Bringing game to an elder is a traditional sign of respect, and with my action I had, quite by chance, come to ask my question in the correct manner. That night I began to record a full cycle of tribal myths.

While trust naturally grows slowly, very often a relationship pivots about a single moment, at which time one can proceed only by instinct and inspiration. Andrew Weil, also an ethnobotanist and writer, told me a story from his time among the Yaqui, a notoriously tough and belligerent tribe of northern Mexico. The Yaqui dances are horrendous ordeals, with the endurance of the participants tested by a week of nonstop line dancing. Andrew is one of the few whites to have been invited to participate. The first time he went he was approached by a tough, extremely proud, and apparently antagonistic Yaqui. They squared off, and the Indian pounded his enormous chest as he bellowed "*Soy Indio*"—"I am an Indio." *Indio* in Spanish is pejorative, so the fellow was essentially saying, "I'm a spick and what are you going to do about it?" Andrew, who happens to be Jewish, suddenly smashed his own chest and yelled "*Soy Judeo*." Then the Yaqui answered with "I'm a filthy Indio," and Andrew countered with "I'm a greedy Jew," and so it went until the two of them had exhausted every conceivable derogatory phrase. Then suddenly they collapsed in laughter, and for the rest of the festival Andrew had this man for his companion and mentor. The point is, the Yaqui had in some sense to display his "Yaquiness" to the visiting white, and by returning his own ethnic bravado, exposing its foolishness, Andrew had cleverly deflected a potentially nasty situation and established a true bond.

My challenge, of course, would be to establish a similar bond with Marcel Pierre.

The bar was deserted, but when Max Beauvoir and I entered the hounfour there were three of them, sitting with their backs to the ochre walls of the bagi. Marcel still hid most of his face, offering only the same cold plastic stare. I greeted him. He made a place for me beside him, near a table.

"Well," he said.

"It didn't work," I told him. Beauvoir lit a cigarette and repeated my words in a low voice. I added, "Ten days I waited and nothing happened."

Marcel showed disbelief.

"Your poison is useless," I said. Then I looked around him at his two companions and asked why they bothered to hang out with a charlatan. One of them made a move toward me. Beauvoir ordered him to sit.

Marcel flushed, and then for the first time the words poured out of him. He called me a liar in a dozen ways before moving indignantly into the bagi. Beauvoir dismissed the onlookers. Marcel came back with

a bag containing the same white aspirin bottle he had shown us the first
day. As he crossed in front of me I grabbed the bag from his hand, tore
off the top of the bottle, and, with my hands above his line of sight,
pretended to pour the brown powder onto my hand. It did not touch
my skin, but Marcel was certain that it had. I pretended to examine the
poison and then to return it to the bottle. I replaced the top and, as I
passed it back to him, cleaned my left hand on the leg of my trousers.

"Sawdust," I said contemptuously.

Marcel fell back, momentarily silent. Flies like huge particles of
dust danced up and down a shaft of light that cut across his face. Looking first at me and then at Beauvoir, he said simply, "He is a dead man."

I got to my feet slowly. "Tell me, then, when will I die?"

Marcel sensed his advantage. "A day, a week, a month, a year. You
shall die from handling that powder."

I drew a breath and felt the hot air sink into my chest. I could only
maintain the ruse. "So what are you trying to tell me? Everyone must
die sometime."

For the first time, Marcel laughed, revealing a row of perfect
teeth. Looking toward Beauvoir he said, "This white of yours is a
brave man," and after a pause, "but he is also very stupid." Only later
would I discover that the small white jar contained the real poison.

Moments later the subject of money rekindled Marcel's fury. It
was one thing to question the quality of his preparations, it was quite
another to demand the return of money that he had already spent. By
now several of his women had slipped back into the hounfour. This
time when he flew into the bagi, it was with a small black vial that he
returned. He placed it delicately, almost reverently, on the table between us. Rage still marked his face, and rivulets of sweat ran across his
hairless brow.

"*Blanc*," he yelled, "you and your kind come a thousand miles to
get my poison. Now you tell me my powder is not good. Why do you
waste my time? Why do you insult us?" He paced and slashed the air
with his arms. His women formed a cordon around him. Then he
stopped.

"If you do not think I make good poison," he said, gesturing toward the table, "drink this and I promise you will not walk out of here
alive."

The circle of faces challenged me. Beauvoir could do nothing.
Marcel came so close I could feel his breath, smell it like a buzzard's.
The silence was unbearable, yet only I could break it.

"Marcel," I said finally, assuming a conciliatory tone, "it is not a question of whether or not you can make good poison. I know that you can. That is why I came a thousand miles to get it. All I am telling you is that what you made me is worthless." I stood up and moved away from the table, rubbing my face with my hands. "You may think that the money I paid you is a lot, but for me it was nothing, for it wasn't my money. To my backers it was so little money that they will not even notice that it's gone. But if you send me back to New York with that useless powder, you will lose the potential to make thousands and thousands of dollars from us in the future."

They seemed stunned. There was a quiver. Then everybody stiffened and held their breath for a silent minute. Some perhaps thought of the money, others may have weighed the insult. Marcel said nothing.

"So you think about that," I said, "and I will be back in the morning." At that Max Beauvoir and I walked out of the hounfour, slowly pushing our way through Marcel's women like men fording a river.

The next morning Marcel greeted Rachel and me at the entrance to his hounfour and invited us into the bagi. It was a small chamber—I could touch a wall in every direction. The air smelled of old newspapers, candle smoke, and earth. Marcel flung open the shutter of the single window, and a shaft of limpid light fell on the altar, turning the colored bottles into jewels. Then he knelt, working a toothpick busily along his gleaming jaw. From beside the altar he lifted a rum bottle and held it to the light. He passed it to me, his gesture a challenge. The bottle was filled with seeds, wood, and other organic debris. The smell was acrid, like rotting garlic. There is an adage in Haiti that warns one never to partake of an open bottle in an unknown temple. I took a drink.

Marcel began to laugh and turned to Rachel incredulously. "How did he know there was nothing in it?"

"He didn't," she replied.

Now Marcel was dumbfounded. "Why is this *blanc* not afraid of me?"

"Because he is afraid of no man," she said flatly. I glanced at her. Though a lie, it was the perfect answer.

At that moment my relationship with Marcel Pierre changed. As we quit the temple I noticed what would turn out to be ingredients of the proper zombi poison drying on the clothesline. Just before we

reached our jeep, Marcel mentioned something to Rachel. She turned
to me uneasily. "He wants you to return tonight alone. He says it is
time to get the poison."

There was no moon, and the clouds blocked the stars. A tremendous thunderstorm had cracked open the dusk sky. Now, shortly after
midnight, by the roadside several miles north of Saint Marc, the dark
clouds remained menacingly on the horizon.

There were five of us—Marcel, myself, his assistant Jean, and two
of his women. We left the road on foot and followed a narrow trail
that crisscrossed up an eroded draw, weaving through spindly, brittle
vegetation. Here, it was pitch-dark. There was one flashlight, but its
fading light was of little use. Marcel carried it in front, stumbling and
laughing in a morbid glee thoroughly seeped in rum. Behind him came
Matilde, her long white dress running behind her in waves as she
walked. I followed, and the other woman, Marie, took my hand. It
wasn't very helpful, for in the darkness she was as clumsy as I, yet I
appreciated the human contact. Jean was the last, and he seemed to
have night vision. He moved slowly, steadily up the draw, his senses
keen, taking in every sound or movement. Slung over his shoulder were
a shovel and a pick.

On a dry knoll, with the hills all around like a finely placed
shroud, the air tasted damp and decayed. The rain was coming. Sheet
lightning flashed in bursts of distant light that revealed shadows on
Marcel's face. In his dark glasses, worn by night as well as day, I saw
reflections of the two women—Marie in red, Matilde in white, her dark
skin glowing out of all that white cloth. Far below, the headlights of
passing cars and trucks skimmed the roofs of the hamlet. The people
there were asleep, while we were about to steal one of their dead.

The tomb was unmarked, just a slight rise in the soil. Jean slipped
away to contact a confederate he had in the community. We waited,
staying low, our arms linked gently. The silence strained my ears. I felt
a flush of fever and fought off the spasms. Jean returned after twenty
minutes, panting, his eyes shining, his lips preserving silence. Marcel
handed him a couple of cigarettes, and Jean carried them into the shadows to the confederate. When Jean struck a match, the bold light momentarily flooded their faces.

The shovel didn't dig, it scraped the compacted soil from the
grave. The pick broke it off in lumps. Deeper and deeper, and from

behind the muffled laughter of the women, like the distant cackle of ravens on the coast at the end of day. From the grave the strong, distinct smell of moist earth.

I held the torch and followed the progress in its narrow beam of light. Some four feet down, the blade of the shovel tore into the reed mat that lined the recesses of the tomb. Beneath the mat were several layers of cotton cloth, the brilliant colors scarcely faded. Then came the hollow thud of steel upon a wooden casket. Jean stopped to cover his face with a red cloth and rub a liniment on all exposed parts of his body. We did the same. Marcel came forward and had us inhale a viscid potion that smelled of ammonia. Cautiously Jean scraped the loose dirt from around the coffin. Leaning as far away from the grave as possible, he reached one arm into the ground and with the pick attempted to pry the coffin from the base of the tomb. The coffin splintered. He stopped, and dug some more with the spade. Finally he crept into the grave, tied a rope to one end of the coffin and hoisted it out of the ground.

It was short, a mere three feet long. Jean cracked the edge of a narrow plank. It took some time for my eyes to grow accustomed to the color of dust and death. Then I felt the horror. I saw a small shrunken head, lips drawn back over tiny yellow teeth, eyes squinting in toward each other. It was a child, a baby girl, her bonnet intact, stiff and gray-brown. As Jean and Marcel carefully placed a large hemp sack over the coffin, I wandered from coffin to tomb. Like a wound, the gaping hole drew me back in strange fascination. Matilde stayed close beside me, stopping me once to wipe my damp brow with the fringe of her skirt. I was dismayed. Bodies decompse rapidly in such a climate: this child could not have been in the grave a month. Jean lifted the coffin onto his head and began to walk down the draw. The others went with him. I came last, following the sugary movements of a whore's hips.

No one paid much attention to us as we unloaded the coffin in front of the Eagle Bar. A few clients leaned over the concrete railing, but the music covered our voices. Jean took the coffin out of the back of the jeep and carried it around to Marcel's bagi. Marcel ordered soft drinks. I bought a couple bottles of rum. I had a few drinks with him, and as I left, I heard Jean working the shovel behind me, burying the coffin in the court of the hounfour. There it would remain until I returned.

Thus we had collected the first and, according to Marcel, most important ingredient of the zombi poison.

I drove south swiftly, my headlights stripping off the final layers of the Haitian night. Beneath the steep slopes that reach the sea near Carrius, with daybreak coming, the pastel sky brightened and luminous clouds revealed great gaps in the sky. Streamers of brilliant light backlit the mountains. Impulsively I made for the sea. Along a pure and virgin shoreline, I felt an irresistible desire to bathe. I shed my clothes on a beach above a fishing village and waded out into the chilling waters. Shapes began to emerge with the dawn—across the water, the shimmering reflections of distant coral atolls, and south along the beach, toward the settlement, the glistening black bodies, piratelike, shouting morning songs. I was glad to be cold. Then I felt the warm breeze fall off the land and caress my cheek. I remembered something the stranger at the Ollofson had said: "Haiti will teach you that good and evil are one. We never confuse them, nor do we keep them apart."

Three days later Marcel led Rachel and me up a broken tract, past a wattle-and-daub house where an ancient woman lived alone, to a draw that opened on a grim wash studded with cacti and brush. Jean was with us, and he and one other assistant carried a metal grill, a cloth sack, and a mortar and pestle. Marcel had a vinyl briefcase, splitting at the seam. We stopped when we reached a small flat, partially shaded by a massive stand of *caotchu*, a wretched succulent with contorted limbs and a viral look. Like everything else in this wasteland, it was sharp and pointed and had sap that burned.

Marcel took his place in the shade, removing his paraphernalia from his briefcase. He placed a thunderstone—a *pierre tonnerre*—in an enamel dish and covered it with a magical potion. Thunderstones are sacred to the vodounist, forged as they are by Sobo and Shango, the spirits of thunder and lightning. The spirit hurls a lightning bolt to the earth, striking a rock outcropping and casting the stone to the valley floor. There it must lie for a year and a day before the houngan may touch it. Despite the divine origins, thunderstones are not uncommon in Haiti; Westerners think of them as pre-Columbian axe heads and attribute their origins to the Arawakan Indians.

Marcel struck a match to the dish, and the potion exploded in flames. Dipping his right hand into it, he set his own skin on fire with the alcohol, then passed the flame to each of us, slapping the joints of our arms and rubbing our flesh vigorously. He then tied satin scarves around our faces to ensure that we did not inhale the dust of the poison. As a final protective measure, he coated all exposed skin with an aromatic oily emulsion.

Earlier in the morning I had watched Jean ease his thick fingers into the coffin, inch his way along the corpse of the child, and close his hand like a vise on the skull. It had collapsed, releasing a chemical scent, foul and repulsive. With great trepidation he had lifted the shattered remains out of the impromptu grave and carefully placed them in a jar. Now, with equal concern, his hand dripping with oil, he took them from the jar and placed them on the ground beside the grill. From his sack he methodically removed the other ingredients. The first two I didn't know—two freshly killed lizards, iridescent and blue. Then he removed the carcass of a large toad that I had seen among other ingredients pinned to the clothesline; dried and flattened it was hardly identifiable, but from its size and a few things he said about it, it had to be *Bufo marinus*, a native of the American tropics, quite common and certainly toxic. Wrapped to the toad's leg was the shriveled remains of what Jean called a sea snake; it looked like some kind of polychaete worm. The toad and the sea worm had apparently been prepared in a particular way: the toad had been placed with the worm in a sealed container overnight before being killed. Jean said that this enraged the toad, increasing the power of its poison. This made sense, for *Bufo marinus* has large glands on the back of its head that secrete some two dozen potent chemicals, a production that increases when it is threatened or irritated.

The plants were easier to identify. One was a species of *Albizzia* known in Haiti as tcha-tcha, and planted as a shade tree throughout the country. The other was *pois gratter*, the itching pea, a species of *Mucuna* that has extremely nasty urticating hairs on its seed pod, hairs that can make you feel that you have slivers of glass under your skin. Jean placed several fruits of both species directly into the mortar. I knew little of the chemistry of these species, but was intrigued that they were both legumes, a family that includes many species with toxic properties. The final ingredients to come out of the bag were a pair of marine fish, one quite innocuous-looking, and the other obviously a puffer fish, not unlike one I had seen on the wall of Marcel's bagi.

My attention was diverted by the young assistant, who had taken a metal grater and begun to grind the tip of a human tibia, collecting

the shavings in a small tin cup. Jean meanwhile had placed the fresh
and dried animals on the grill and was roasting them to an oily consistency before transferring them to the mortar. The bones of the child
stayed on the grill until burned almost to charcoal. Then they too were
placed into the mortar. By the time all the ingredients were ready to be
crushed, the smoke that rose from the vessel was a corrosive yellow.

I glanced at Marcel, somewhat confused by his role. He never
touched the poison nor any of the ingredients. He lay back in the
shade, occasionally shouting instructions, his mind attentive and his
eyes scanning the trail, watching for intruders. One time two young
children came by, and he jumped up to chase them away, shouting in-
sults and threats. Yet the entire time we were in that scrubland there
was a family tending its fields on a nearby slope watching our every
activity. Marcel even exchanged words with them occasionally, shouting across the narrow valley. He seemed to be maintaining a pretense
of secrecy while remaining on constant display, and quite proud of it.
Herein I realized lay the essence of his ambiguous position. Like the
sorcerers in Africa he was despised by all upstanding members of the
community, yet at a more profound level his presence was tolerated
because it was critical to the balance of social and spiritual life. The
bokor and all his apparently maleficent activities were accepted because
they are somehow essential.

But what was Marcel? A bokor, a malevolent sorcerer; or a houngan, a benevolent priest and healer? Beauvoir had pointed out the fal-
lacy of any such distinction. Marcel, of course, was both, yet himself
neither evil nor good. As bokor he might serve the darkness, as a houngan the side of light. And like all of us he was capable of serving both.
The vodoun religion had explicitly recognized this dichotomy, and
had in fact institutionalized it. This was why Marcel's *presence* was
critical. Without his spiritual direction our activities had a completely
ambivalent potential for good or evil. It was he who was ultimately
responsible.

There seemed little doubt which force Marcel now chose to embrace. *It had not been his desire to go into the graveyard, it had been
mine.* I had commissioned the poison, for which the bones were necessary. That night—and now in this barren land where creepers wove nets
over stones, where plants had leaves that breathed by night and collapsed to the touch—it was Marcel who assured our safety. And so he
as houngan had no contact with the poison. Such a destructive force

had to be prepared by Jean, who was neither an apprentice nor an assistant but rather a physical support. When the BBC and others described Marcel Pierre as the incarnation of evil, they had missed the point completely.

These intuitions of mine became even stronger when Marcel began to sing. By now Jean was pounding the ingredients in the mortar, and the steady thud of the pestle laid down Marcel's rhythm. Then the young assistant took up a pair of ordinary stones, struck them together, and we had percussion. Then Rachel joined in—for she knew all the songs—and her soft voice rose highest, flowing back and forth, teasing in its beauty. Marcel's entire body melted into the rhythm, and it seemed that at any moment he might become possessed. His broad smile, his radiant participation in the songs—here in the middle of making this poison—it was his joy, this pure unadulterated joy that made me think that somehow this could not be evil.

Just as the poison neared completion, as Jean sifted the residue from the mortar, a quite accidental exchange took place that in the end seemed to have a profound impact on Marcel. I say accidental, though in truth I was becoming skeptical of things accidental, of chance and coincidence. In Haiti nothing seemed to happen as it should, but little occurred by chance.

At any rate, I was wearing a knife on my belt, and Marcel asked if he could have it. The knife was important to me—I had traded for it some years before at the headwaters of the Amazon on the Rio Apurimac in Peru. I told Marcel that it was my most valued possession and impossible to part with, that it had been given to me following the completion of one of my people's most important rituals. This was untrue, but the impetus of the lie carried me into a true account in which the knife was promptly forgotten.

First I tried to share with him a new notion of space. I spoke of mountainous valleys near my home in northern Canada, valleys larger than all of Haiti and totally uninhabited. I described moving through lands where space yielded in every direction to the infinite. I spoke of tundra vegetation at one's feet, a cornucopia of color and sound, of whistles and birdcalls, of rust and ginger splashed onto a canvas that stretched to the horizon—and there, forests of mountains wrapped in icefields, seething masses of rock and ice in an ocean of clouds. I explained that between these two extremes, the minute flora and the gargantuan mountains, there was a complete dearth of man-sized objects.

I tried to make Marcel envision a land where men were insignificant. It was perhaps the most difficult thing for him to understand. Then I spoke of temperatures, of lakes solid with ice, of damp clothing left out overnight and cracking the next morning like a stick. I described hunting animals, moose and caribou, speaking of the number of pounds of meat that each yielded. I spoke of wolves and bears, myths and legends passed on to me by the old hunting guides. Then I told him of the vision quest.

I explained how as a younger man I had been instructed to climb the highest peak in the valley, while the old Gitksan Indian waited below. I tried to carry Marcel up that mountain, describing in great detail the route, the steep scree and the ledges alive with goat, the dizzying exposures and the whirling landscapes of waterfalls and rock, spruce forests, and glaciers. I had him build with me, stone by stone, a rock cairn. I had him watch me as I sat alone on the summit, without food or water, until the animal came, what kind of animal I could never say. That animal arrives, I told Marcel, not by chance but because it was fated to become one's protector—a spiritual guardian that might be called upon five or six times over the course of my lifetime. And so I explained to Marcel that I had my animal, and that was the reason I was not afraid of him. That was why I feared no man.

Once this came clear to Marcel, he became visibly excited. Quite by chance the vision quest I had experienced bore striking parallels to fundamental features of vodoun initiation. The *hounsis canzo*, or initiate, enters a week of seclusion in which he or she suffers a particular diet and rigorous prescribed activities, all supervised by an elder. At the end of a week the individual receives a spiritual name and enters the path of the loa, the divine horsemen. And so I finally made sense to Marcel. Later that day Rachel overheard him explaining to several of his people that this *blanc* was unlike the others because he had been initiated. It seemed a ridiculous way to go about things, he had told them, but that was the way people acted in the impossible land of Canada.

A pair of Marcel's women lounged on the front steps of the Eagle Bar. His was an ugly trade, made worse by the innocence of the women at midday—their hair bound in curlers, their nails gleaming in fresh reds and purples—yet there remained something guiltless about Marcel's establishment. One sensed that the Haitian men with their astonishing collection of wives, mistresses, and *mamans petites* had no shortage of outlets for their desire. They came to Marcel's out of curiosity, sometimes to slake their flesh but most often just to gather. Behind the facade of bar and brothel the place had the atmosphere of a neighborhood club, informal and intimate.

Inside the bar Marcel and I celebrated our newfound trust by sharing a plate of rice and beans. Gleefully he explained in minute detail how he had bluffed me, recounting like a master storyteller each moment of our first encounter. He also instructed me in the application of the poison. As Kline had suggested, it could be spread in the form of a cross on the threshold of the victim's doorway. But Marcel also said that it could be placed inside someone's shoe, or down his or her back. This was the first indication that the poison might be applied directly to the intended victim. It made sense, of course, given my suspicion that there was no way any poison could get through the callused feet of a Haitian peasant. Moreover, placed on the ground at the entrance of the hut, it would presumably affect everyone who stepped on it, not just the intended victim.

With my confidence in Marcel reasonably established, my attention turned to the reputed antidote. Kline had mentioned several reports suggesting that a chemical antidote was administered to the zombi victim in the graveyard at the time of his resurrection. When I brought up the subject with Marcel, he remained equivocal. It was strictly the power of the bokor that revived the dormant zombi, he claimed. With two assistants the bokor would enter the cemetery, approach the grave, and call out the victim's name. The zombi would come out of the ground unaided, be promptly beaten and bound, then led away into the night. His description coincided closely with the account of Clairvius Narcisse. Marcel went on to indicate that there was, however, a preparation that if used properly completely counteracted the effects of the poison. When I asked if he would be able to prepare it for us, he looked momentarily bewildered. Naturally, he replied, one would never make the poison without making the antidote. Marcel glanced toward Rachel as if I were the stupidest man alive.

That afternoon Jean dug up the young girl once more. After carefully placing the jars of poison in the coffin, he covered it over and retired to a corner of the hounfour. His work was momentarily finished. The poison would remain with the corpse for three days. Not surprisingly, it was Marcel, not Jean, who then mixed the ingredients of the antidote. He began by placing in a different and larger mortar several handfuls of dried or fresh leaves of six plants: aloe (*Aloe vera*), guaiac (*Guaiacum officinale*), cèdre (*Cedrela odorata*), bois ca-ca (*Capparis cynophyllophora*), bois chandelle (*Amyris maritima*) and cadavre gâté (*Capparis* sp.). This plant material was ground with a quarter-ounce of rock salt, then added to an enamel basin containing ten crushed moth-balls, a cup of seawater, several ounces of clairin or cane alcohol, a bottle of perfume, and a quarter-liter of a solution purchased from the local apothecary and known as *magie noire*—black magic. Additional in-gredients included ground human bones, shavings from a mule's tibia and a dog's skull, various colored and magically named talcs, ground match heads and sulphur powder. It was straightforward procedure, devoid of ritual or danger. The end product was a green liquid with a strong ammonia scent, similar to the substance that Marcel had been rubbing on us all along.

Below ground in the open court of the hounfour rested the child with the glass jars of poison cradled in her lap. Above ground one of the assistants placed lit candles at either end of the buried coffin, then traced in cornmeal a cabalistic design that bound the child's new grave to the altar of the bagi. On the surface of the court he traced a second coffin and dissected it into fourths by drawing a cross. In each quadrant he drew the symbol of a spirit of the dead. Marcel poured the antidote into a rum bottle and placed it upright over the grave, its base buried in the earth, its mouth pointing to the sky.

Curiously, while the poison contained many ingredients with known pharmacological activity, the antidote was decidedly unin-teresting from a chemical point of view. Most of the ingredients were either chemically inert or used in insufficient quantity. More impor-tantly, the way that the antidote was applied strongly suggested that it had little to do with the actual raising of a zombi. It was only once somebody knew that he or she had become a victim of the poison that the antidote was administered, and it was applied simply as a topical rub. The antidote wasn't intended to revive the victim from the dead; it merely prevented him or her from ultimately succumbing to the poi-son. And it did so according to a particular timetable. If a victim knew that he or she had been exposed to the poison no longer than fifteen days, they could simply administer the antidote. If, however, one had been subjected to the poison for more than fifteen days, the antidote had to be augmented by an elaborate ceremony in which the victim was symbolically buried alive. In other words, for severe cases it was not the antidote but a body of ritual and belief that was responsible for survival. A pharmacologically active antidote might still be discovered, but it was certainly not the one concocted by Marcel Pierre. Nor did he deny this. For him, the antidote was the power of his own magic.

Interpreting this new information made clearer what a vodounist considers to be a poison. Kline, the BBC, and others had obviously taken reports of an antidote too literally, assuming that it had to be a substance used to resurrect the zombi from apparent death. Such straightforward cause and effect, which I had tried to answer with the Calabar hypothesis, had seemed reasonable. It was logical and linear—just what the Haitian spirit realm is not. Out of curiosity I asked Marcel to name the greatest poison. There was no doubt, he replied, that far more deadly than the preparation we had made, more dangerous than even human remains, was a simple lime, properly prepared by the bokor. According to Marcel, and many other houngan I later asked, if a bokor cuts a lime transversely while it is still on the tree, the half that remains on the limb overnight becomes the most virulent of poisons. The other half taken into the temple becomes its equally potent antidote. The lesson was clear. The lime that is left on the tree remains in the realm of nature—uncivilized, threatening, poisonous. The other half, taken into the abode of the religious sanctuary, is tamed and humanized, and thus becomes profoundly curative. Apparently just as man himself has an ambivalent potential for good or evil, so do objects; in the case of the lime it is the intervention of man alone that may release its latent promise. For the vodounist there seem to be no absolutes. Only the houngan embodies all the cosmic forces and maintains their balance. So it was man who ruled the Haitian worldview, and it was the power of man that treated the poison victim. Similarly, I concluded, it would be man and not a poison that created a zombi.

Having tracked down the poison and its supposed antidote, I had very little time left before I was due back in Cambridge. But in those last days a curious event took place. At the time it seemed significant. One afternoon in Saint Marc Max Beauvoir and I were approached by two men in peculiar uniforms and flat hats like those worn by the U.S. Marines when they occupied Haiti in the early years of this century. The men were members of a paramilitary cavalry group from the town of Desdunnes in the Artibonite Valley. This reputedly placed them as descendants of the legendary highwaymen who, dressed in loincloths and brandishing swords that deflected bullets, terrorized the caravans during the French colonial era. Much later during the American occupation, many of the men of Desdunnes joined the *cako*, the resistance fighters who waged guerrilla war from sanctuaries deep in the mountains of northern Haiti. Belief still has it that the leader of the U.S. Marine command was zombified by the cako. Today the residents of Desdunnes remain among the most fiercely independent people of Haiti.

The two men had heard that Beauvoir had influence in the capital, and they wanted to know if he might be able to arrange for them to demonstrate their horsemanship for the president. They offered to put on a private demonstration for Beauvoir the following week in Desdunnes.

The day before I was scheduled to leave Haiti, Max Beauvoir, Rachel, and I arrived in Desdunnes, finding more than fifty mounted men waiting beside their commander's house. By American standards the horses were small, but all appeared well trained, and the men put on an impressive display. At midmorning, as the horsemen broke off and gathered by the plaza to drink with us and the commander, I happened to mention that I enjoyed horses. The commander took that casual comment as a signal that I wanted to ride with his group. Within minutes I was atop an exhausted horse. The commander asked Beauvoir if they ought to lead the horse by the halter, but by then a couple of the men and I were trotting out of the plaza. Then, as they realized I could indeed ride, the men broke into a gallop, and those on foot began to shout, "*Savandier.*" At that point, as I would later learn, the commander turned to Beauvoir and said, "A man who rides like that was born on the savanna. So, my friend, you come to have a look at us, and you bring your own rider." Then after a pause he went on, "Let's have a look at him. We have a horse that doesn't dance, but it does run."

We tethered our horses, and after a few drinks the rhythm of the afternoon set in. We had been sitting for about an hour when two men returned with a fresh brown mare.

The town was by then in a frenzy. It was Easter, and the Rara bands had swarmed out of the temples to celebrate the end of Lent. Their processions wove through the streets, swirling past one another, invading gardens and homes, absorbing idlers and growing longer and longer tails of dancers. From a distance they could be taken as hallucinations, except for the music—a single four-note song. The bass line came from four long bamboo tubes. Tin cans transformed into trumpets and trombones created a glittering horn section; rubber hose transformed into tubas created another. Percussion in the hands of a Haitian is anything that knocks—two sticks, a hubcap, a hammer-and-leaf spring from a truck.

The bandleader was a malevolent jester, somewhat androgynous. The others in the lead looked like the Queen of Hearts, only more lascivious, in long satin dresses sporting lewd bustlines. All of them were men. Sweeping across the head of the procession was a menacing figure, wielding a sisal whip, flailing at the crowd. But it was a symbolic display, and no one was hurt. Nevertheless Rara remains somehow intimidating and subversive. It is an amazing sexual inversion and an extraordinary triumph of the spirit. No wonder that by political decree the bands are permanently forbidden from entering the major cities.

So, with a gallery of dusty-faced peasants interwoven by the brilliant, swirling colors of the Rara bands, I got onto the horse. Two men on foot released the halter, and four others on horseback led me out of the plaza to the main entrance of town. As our trot broke into a canter, two of the other horsemen fell back. Then we began to run, and I found myself with the remaining horseman in an unanticipated contest. His challenge thinly disguised, he led me on a wide circuit about town, past the thatched huts and across the crowded plazas. At one point when I veered onto the wrong trail, the horseman paused and turned back to me, laughing. Then, as I drew abreast again, he loosened his rein and let out a great howl. The race was on in earnest. Women grabbed their children or chased away the chickens, and the town's main thoroughfare became a storm of dust. As we circled the mapou tree in the central plaza, my horse entered the lead. The peasants began to shout, "*Blanc, blanc*" as froth from the neck of my horse struck my face.

Although I crossed the line a few paces ahead of my companion, it clearly mattered less who won than that the race had actually taken place. With obvious delight the commander took my reins as I dismounted, then led me into the courtyard of his house. Surrounded by the rest of his horsemen, we posed for the local photographer. Finally he led me into the house to join the Beauvoirs for lunch.

Only later would Max Beauvoir tell me that the commander was a president of a secret society, that I was now considered a member of his horse troop, and that the meal had been prepared and served by three queens of the secret society.

That afternoon we left Desdunnes and continued north to an important ceremonial center in the Artibonite Valley. There, once again by chance, I came across an important clue, yet it was a discovery that, like so many made during this first phase of my investigation, only deepened the mystery. At dusk, in the fields behind this most sacred of sites, I stumbled upon an entire field of planted datura. The next day I left Haiti and returned to the United States.

PART TWO

Interlude at Harvard

7

Columns on a Blackboard

ON EASTER SUNDAY I passed through United States Immigration at Kennedy Airport in New York carrying a kaleidoscopic Haitian suitcase constructed from surplus soft drink cans. The specimens inside included lizards, a polychaete worm, two marine fish, and several tarantulas—all preserved in alcohol—as well as several bags of dried plant material. Two rum bottles contained the antidote, while the poison itself was in a glass jar wrapped in red satin cloth. There was also a dried toad, several seed necklaces, a dozen unidentified powders, and two vodoun *wangas,* or protective charms. Two human tibia and a skull were at the bottom of the case. A cardboard box was full of herbarium specimens, and concealed inside in a duffle was a live specimen of the bouga toad. The customs agent opened the cardboard box, took a quick look, and told me he didn't even want to hear about it. He never saw the toad.

There was no one in when I rang Nathan Kline from the airport, so I left word on his answering machine and caught the next plane for Boston, arriving in Cambridge slightly after dusk. I found the Botanical

Museum deserted. I was tired, and when I entered my office I placed
the tin suitcase on my desk and, without turning on the lights, lay
back in a hammock I had strung across a corner. It was amusing to
look at that colorful case so symbolic of an entire nation. Haiti, it is
said, is the place to discover how much can be done with little. Tires
are turned into shoes, tin cans into trombones, mud and thatch into
lovely, elegant cottages. Material goods being so scarce, the Haitian
adorns his world with imagination. Yet if there was anything silly about
the suitcase, there certainly was not about its rank contents. If the
mystery of the zombi phenomenon was to be solved, these specimens
were the most important clues. Without them, there was nothing concrete. With them, I could take full advantage of the resources of the
university.

I stood up and emptied the case, placing the specimens in a long
row across the back of my desk. Then I took a piece of chalk and drew
a column at either end of a large blackboard that covered much of one
wall. On the left I listed the ingredients in Marcel's poison: human
remains, the two plants, the sea worm, toad, lizards, and fish. On the
right I wrote the medical symptoms of Clairvius Narcisse at the time
of his death: pulmonary edema, digestive troubles with vomiting, pronounced respiratory difficulties, uremia, hypothermia, rapid loss of
weight, and hypertension. After a moment's recollection, I added
cyanosis and paresthesia, for both Narcisse and his sister had mentioned
that his skin had turned blue, and that he felt tingling sensations all
over his body. In between the two lists was a very large blank space.

Early the next morning I passed through the dark corridors of
the Museum of Comparative Zoology and deposited the animal specimens with the museum's various specialists for identification. Then I
returned to the Botanical Museum to have a look at the plants. I had
three initial questions. Did the plants have pharmacologically active
compounds that might cause a dramatic decrease in metabolic rate?
If so, were the strength and concentration of the compounds sufficient
to be effective in the doses used by Marcel Pierre? And finally, were
they topically active, or did they have to be ingested? All three were
tempered by an important consideration. Very often in folk preparations, different chemicals in relatively small concentrations effectively
potentiate each other, producing a powerful synergistic effect—a bio-
chemical version of the whole being greater than the sum of the parts.
Thus my primary concern was to demonstrate whether or not any of

the plant or animal ingredients contained interesting chemical compounds. Once the specimens were identified, I would find most of this preliminary information in the library.

Late that afternoon the museum secretary interrupted me with a message from Kline asking me to return as soon as possible to New York.

The door to the apartment was open. I went in, stood rather tentatively for a moment in the alcove, then wandered into the living room. The decor was unchanged, but the late afternoon sun left the air hot and compressed and tinged every object with color.

"You seem to be all right." I turned around, and Kline's daughter Marna was behind me, smiling and leaning against the doorway to the kitchen. "Welcome back."

"Hello, Marna. Sorry if I'm barging in."

"It's fine. Father told me you'd be coming. How was Haiti?"

"It's a good place."

"And the work?"

"Okay."

"Was he right?"

"Your father?"

She nodded. "No, don't tell me, don't say anything. Wait until he gets here. Can I pour you a drink?"

"That'd be nice."

"Some rum, or have you had it?"

"I'll take a whiskey."

Marna went back into the kitchen to get some ice and then walked over to the bar. "He was worried, you know."

"Your father? Why?"

"Weren't you a week late getting back?"

"I don't think so."

"He expected you several days ago."

"There were a few delays," I said. "Where's your father now?"

"Oh, I'm sorry. He just stepped out. He should be right back. Bo Holmstedt is with him. Do you know him?"

"He is? Yes I do." This was a surprise. Bo Holmstedt is a professor at the Karolinska Institute in Stockholm and one of the world's foremost toxicologists. He and Schultes were good friends and had collaborated on a number of projects in the Amazon.

Marna and I were having a second drink when her father and Holmstedt returned. Nathan Kline seemed a different man from our first meeting—warm, even affectionate as he greeted me.

"There you are. All in one piece, I'm glad to see. You know Bo Holmstedt?"

"Professor."

"Hello, Wade."

"Bo was on his way to the airport, but I've persuaded him to wait until he hears what you've got to tell us."

"Now, Nathan, don't put the lad on the spot," Holmstedt said. "Marna, haven't you got a quick drink for an old sot?" He spoke with the accent of a Scandinavian educated in Britain. An elderly man, short and heavyset, he was dressed conservatively in gray flannels and a blazer. We chatted informally until all were comfortably settled, and then I began my account. In a necessarily anecdotal way, and responding to frequent questions, I shared my instincts for the country, its history and unique social structure, stressing repeatedly the importance of the vodoun religion as the axis around which so much of Haitian life revolves. Then I recounted the events that led up to my obtaining the preparation from Marcel Pierre. Finally, the conversation turned to the poison itself.

"I haven't got all the determinations yet on the animals, but I have identified the plants. One is a liana known to the Haitians as pois gratter, the 'itching pea.' It's *Mucuna pruriens*, and like just about everything else in the genus the fruits are covered with vicious urticating hairs."

"Anything on its chemistry?" Kline asked.

"Not much. I spoke with Professor Schultes, and he seems to think the seeds are psychoactive. He's also seen it used medicinally in Colombia to treat cholera and internal parasites."

Holmstedt joined in. "There is a species of *Mucuna* called . . . Damn it, what is it? Yes, *flagellipes*. It's used in central Africa as an arrow poison. Has something in it not unlike physostigmine. You know, of course, the literature on the Calabar bean?"

It was a good thing I did, for Holmstedt had written a large portion of it. I reviewed briefly the hypothesis I had come up with, and explained that there was no evidence that the plant had reached Haiti.

"A good hypothesis is never wasted. Now what else have you got?"

"*Albizzia lebbeck.* The Haitians call it tcha-tcha. It's a native of West Africa introduced into the Caribbean some years ago as an ornamental shade tree."

"You may be onto something there. What did you find out about it?"

"The bark and seedpods contain saponins. In small concentrations quite effective as a vermifuge. But dosage is critical."

"Isn't it always?" said Kline.

"Some West African tribes use the plant as an insecticide or fish poison," I said.

"How's it work on fish?" Marna asked.

"They put the crushed seeds in shallow bodies of water. The saponins act on the gills to interfere with breathing. The fish suffocate, float to the surface, and the people gather them up. Doesn't affect the meat."

"You will find that killing a mammal with saponins is somewhat more complicated," Holmstedt said. "But it can be done. In East Africa, the roots of *Albizzia versicolor* make a bloody fine arrow poison. But it's no good putting it in someone's food. Saponins aren't absorbed by the intestines. You've got to get the stuff into the blood."

"But isn't that what they say they do in Haiti?" Marna asked.

"Yes," I replied, "through the skin. Professor Holmstedt, what are the symptoms of saponin poisoning?"

"In sufficient dosage, nausea, vomiting, eventually excessive secretions into the respiratory passages. In simpler terms, the victim drowns in his own fluids."

"With pulmonary edema as an interim result?" I asked.

"Yes, of course." It was the condition that headed my list of Clairvius Narcisse's symptoms at the time of his death.

"You will find that *Albizzia* has one more secret," Holmstedt went on. "Many species of the genus contain a special class of compounds known as sapotoxins that are absorbed by the intestines. They work in a rather nasty fashion by interfering with cellular respiration throughout the body. They kill you by actually weakening your every cell. Incidentally, the Efik traders of Old Calabar used the bark of *Albizzia zygia* in a potion known as *ibok usiak owo*, which means in their language 'a medicine for mentioning people'—a sort of native truth serum, I would say. They administer it orally as an ordeal poison. You see, Wade, you come back full circle to your hypothesis."

"Have you got anything yet on these lizards and toads and whatever else they put in?" Kline asked.

"Nothing encouraging," I replied. "There's a polychaete worm that Marcel sequesters with the toad. It had bristles on it that some say inflict a mild paralysis. They may be venomous, but the reports are vague. The herpetologists up at the museum knew the lizards right away. Neither is known to be poisonous. In Dominica, they say one of them can make the hair fall out and turn your skin green. But the people eat it anyway. Another species in a related genus plays havoc with housecats in Florida, but it doesn't kill them. But it wasn't a lizard that took down Narcisse, that's for sure."

"What about the toad?" Holmstedt asked.

"*Bufo marinus.*"

"You're certain."

"It's unmistakable. The people at Herpetology confirmed it."

Holmstedt paused to consider. "Quite," he said finally, and looking straight at me, "Wade, I'd say you're onto something."

The next morning when I returned to Cambridge, there was no new information for me. The first botanical determinations did show that Marcel Pierre was exploiting plants with both pharmacologically active compounds and proven African connections, but isolated reports drawn from over half a continent explained very little. The Calabar hypothesis, despite Holmstedt's words of encouragement, seemed to have led me nowhere. I had found no evidence of the Calabar bean in Haiti, and while datura had been present, its connection to the zombi phenomenon remained uncertain.

It was discouraging, but my luck would change by the end of the day. The more I read about the large bouga toad, the ingredient that had so caught Holmstedt's interest, the brighter my mood became. I learned that the paratoid glands on its back are virtual reservoirs of toxic compounds.

Although *Bufo marinus* is a native of the New World, it apparently reached Europe very soon after the voyages of Columbus, and there it was well received by nations long familiar with toad venom. Europeans believed that toads derived their poisons from the earth by eating mushrooms (hence the English name *toadstool*). As early as Roman times, women used toads to poison their husbands. Soldiers in the Middle Ages believed that a discreet means of wounding an enemy was to rub his skin with the secretions of *Bufo vulgaris,* the common European species. And not long after *Bufo marinus* reached the Old World, poisoners found that by placing the toad in boiling olive oil, the secretions of the glands could be easily skimmed off the surface. In Italy early in the sixteenth century, poisoners devised sophisticated processes for extracting toad toxins into salt, which could then be sprinkled on the intended victim's food. In fact, so highly regarded was the toxicity of toad venom that at the beginning of the eighteenth century it was actually added to explosive shells. Presumably the commanders felt that if the cannon did not kill their enemies, the toad toxins would.

The toxic properties of toads had certainly not been overlooked by the natives of the Americas. The Choco Indians of western Colombia, for example, learned to milk poisonous toads by placing them in bamboo tubes suspended over open flames. The heat caused the creature to exude a yellow liquid, which dripped into a small ceramic vessel where it coagulated into the proper consistency of curare, the arrow poison. According to an early and perhaps exaggerated report, these preparations were extremely potent; a deer struck by an arrow survived two to four minutes, a jaguar perhaps ten. Of course, arrow poisons work in many different ways. Those based on the lianas of the northwest Amazon act as muscle relaxants, causing death by asphyxiation. The skin of *Bufo marinus,* on the other hand, contains chemical substances resembling the strongest of the African arrow poisons. These latter are derived from a plant, *Strophanthus kombe,* and they act in a quite different way. The active principle is a chemical called ouabain, and it is a powerful stimulant of the muscles of the heart. In moderate dosage ouabain is used today to treat emergency heart failure; in excessive doses it makes the heart go crazy, pumping wildly until it collapses.

Not surprisingly, European physicians incorporated toad venom into their materia medica at a very early point, and in fact powdered toad remained a prominent therapeutic agent well through the eighteenth century. But as in so many things, the Chinese were far ahead. For centuries they had been forming the venom into round, smooth dark disks, which they named *ch'an su,* or "toad venom." According to the *Pentsao Kang Mu,* a famous herbal written at the end of the sixteenth century, the venom was used to treat toothache, canker sores, inflammations of the sinus, and bleeding of the gums. Taken orally as a pill, it was said to break up the common cold.

From this list of rather mundane afflictions, it is difficult to appreciate that the Chinese were dealing with an extremely toxic preparation. When ch'an su was analyzed early in this century it was found to contain among other active principles two powerful heart stimulants known as bufogenin and bufotoxin.

Powerful is perhaps an understatement. Virtually any contemporary victim of heart disease depends on a daily dose of digitalis, a drug extracted from the common European foxglove, a plant used as a cardiotonic in England since the tenth century. Bufotoxin and bufogenin have been found to be fifty to one hundred times more potent than digitalis. What does this mean? When a cat, in one experiment, was injected with as little as one-fiftieth of a gram of crude ch'an su, its blood pressure tripled almost immediately, and then it collapsed following massive heart failure. If a human responded in the same way, this would mean that as little as half a gram of dried venom applied intravenously would do similar punishment to a 150-pound man. Applied topically, one would at least expect a rapid increase in blood pressure. A quick glance at my list of Narcisse's symptoms allowed me to draw a line between *Bufo marinus* and hypertension.

Given the toxicity of these compounds it is perhaps difficult to appreciate a controversy that has developed in recent years over whether or not *Bufo marinus* was used as a hallucinogen by New World Indians. The problem is that the glands of the toad secrete another chemical known as bufotenine, a compound that is found in a hallucinogenic snuff made from a plant by Indians of the upper Orinoco in Venezuela. In Central America the toad seems to be a prominent feature of Mayan iconography, and at one Postclassic Mayan site, in Cozumel, Mexico, an archaeologist found that virtually all amphibian remains were *Bufo marinus*. This report coincided well with an earlier and similar discovery at San Lorenzo that led one prominent archaeologist to suggest that the Olmec civilization used *Bufo marinus* as a narcotic. As it turns out, however, *Bufo* remains the dominant amphibian component of middens throughout many parts of Central America, leading other archaeologists to believe that pre-Columbian Indians didn't get high from the toad, but ate it only after carefully cutting away the skin and paratoid glands. Today in Peru, for example, the toad is commonly eaten in this manner by the Campa, a tribe of the upper Amazon. This seems to be the most reasonable idea, especially when one considers what kind of intoxication the toad would offer. It is unlikely that the Maya would have been interested in selectively poisoning vast numbers of their priesthood, who presumably would have been the ones taking the drug. Only if some very complex process had been developed that selectively neutralized its toxic compounds could *Bufo marinus* have been much of a ritual hallucinogen.

Still, such a process was not inconceivable. Several years ago a colleague of Professor Schultes dispatched an intrepid young anthropologist by the name of Timothy Knab to search the backcountry of Mexico for a contemporary cult that might have preserved the ancient knowledge. After months of effort Knab finally located an old curandero in the mountains of southern Vera Cruz who knew the formula of a preparation that had not actually been used by his people for fifty years. The old man ground the glands of ten toads into a thick paste, to which he added limewater and the ashes of certain plants. The mixture was boiled all night, or until it no longer smelled foul, and then was added to corn beer and filtered through palm fiber. The liquid was mixed into cornmeal and then placed in the sun for several days to ferment. Finally, the mixture was heated to evaporate the remaining liquid, and the resulting hardened dough was hidden deep in the forest.

Although Knab had persuaded the curandero to prepare the drug, under absolutely no conditions would the recalcitrant old man actually sample it. Only very reluctantly did he consent to give a dose to Knab. From what happened it appears that he knew something the anthropologist did not. Knab's intoxication was marked by sensations of fire and heat, convulsive muscle spasms, a pounding headache, terrifying hallucinations, and delirium. For six hours he lay immobilized in a specially excavated depression in front of the curandero's fire.

Knab never did find out whether or not the preparation actually neutralized any of the most toxic compounds found in the glands of the toad. But even if it had, it seems that bufotenine alone would be enough to ruin any experience. In the late 1950s Howard Fabing, a medical doctor, obtained permission to inject bufotenine intravenously into a number of inmates at the Ohio State Penitentiary. In the mildest dose an inmate complained of a prickling sensation in his face, nausea, and slight difficulty in breathing. In a higher dose these symptoms became more pronounced, and the face and lips became purplish. The final doses caused mild hallucinations and delirium, and the skin turned the color of an eggplant, indicating that the drug was keeping oxygen from getting into the blood. Further experiments led this audacious physician to conclude that the symptoms produced by bufotenine coincided curiously with the conditions of the *berserkus* of Norse legend. Our expression *going berserk* pays homage to these warriors who, according to Fabing, ingested a psychoactive substance that put them into a state of frenzied rage, reckless courage, and enhanced physical strength. True or not, it was worth considering his conclusions in light of the zombi investigation. Fabing's descriptions of his experimental subjects closely matched those of the zombis when they first come out of the ground. Marcel Pierre had suggested that as many as three men might be required to subdue the zombi; and Narcisse had mentioned that he had been beaten and bound as soon as he came out of the ground.

Again I considered Narcisse's symptoms at the time of death, and was gratified to find that they included both cyanosis (the bluing of the face) and paresthesia (the tingling sensations).

By the end of the day some interesting patterns were appearing on the blackboard. With some confidence I had eliminated the two species of lizards. The polychaete worm was more problematic, as there was very little information on the nature of its reputed toxin. The two reports I had found had been vague. My tendency was to accept the interpretation of the vodounists themselves, who had suggested that the role of the worm was simply to agitate the toad, and thus increase the quantity of toxic secretions. One of the plants did have recognized toxins that were topically active and caused pulmonary edema. And of course the toad had a slew of pharmacologically active compounds, all topically active, some of which might cause severe hypertension, cyanosis, and paresthesia as well as behavioral changes marked by delirium and a confused state of sham rage.

Yet clearly it was not enough simply to evaluate chemical properties of the ingredients and compare them with the symptoms of Narcisse. *Albizzia lebbeck*, for example, can cause pulmonary edema, but so might a dozen other substances. Missing from this data was any evidence of a potent compound that might bring about the most critical requisite condition—a profound reduction in metabolic rate that would actually cause the victim to appear dead.

By the middle of the week I had still not heard from the fish experts, so I decided to drop by their lab to see what was going on. I found the man who was working on my specimens in a dark corner of their basement catacomb, staring down the tiny mouth of an incredibly ugly creature. I think he was counting teeth.

"Got anything on those Haitian fish yet?" He looked up from his specimen, struggling both to remember who I was and to refocus his attention on something of human dimensions.

"Ah, yes. Haitian fish. Nasty little beasts." He slipped into the back room and returned with my specimens. "Schultes have you working for the CIA, or what?" He laughed out loud.

"How do you mean?"

"I mean the puffers." He rattled off a list of scientific names that meant nothing to me.

"What do they have in them?"

"Good Lord, I thought you people were drug experts. Not very up on your literature, either." I must have looked confused. "James Bond. Last scene in *From Russia with Love*, one of the great moments in ichthyotoxicology. British agent double-oh-seven utterly helpless, paralyzed and unconscious after a minute wound from a hidden knife." He stood up and perused his bookshelf, somehow managing to look scholarly even as he pulled the small paperback from between the thick rows of anonymous journals.

"Knew it was here somewhere. Here you go." He quoted: " 'The boot with its tiny steel tongue flashed out. Bond felt a sharp pain in his right calf. . . . Numbness was creeping up Bond's body. . . . Breathing became difficult. . . . Bond pivoted slowly on his heel and crashed to the wine-red floor.' " He returned the paperback to the bookshelf. "Double-oh-seven never had a chance," he lamented. "Terribly clever of Fleming, too. You have to read the next book to find out. The blade was poisoned with tetrodotoxin," he confided. "He tells you in the first chapter of *Dr. No.*"

"What is it?"

"A nerve toxin," he replied, "and there is nothing stronger."

It didn't take me long to realize that the original hunch of Kline and Lehman had proved correct: the zombi poison included one of the most toxic substances known from nature. Marcel had recognized two varieties of fish—the *fou-fou*, which was *Diodon hystrix*, and the crapaud de mer, or sea toad, which was *Sphoeroides testudineus*. In English we know these as blowfish or puffer fish because of their ability to swallow large amounts of water when threatened, and thus assume a globular shape, making it more difficult for their predators to swallow them. One would hardly think such a passive defensive mechanism necessary. Both creatures belong to a large pan-tropical order of fish, many of which have tetrodotoxin in their skin, liver, ovaries, and intestines. This deadly neurotoxin is one of the most poisonous nonprotein substances known. Laboratory studies have shown it to be 160,000 times more potent than cocaine. As a poison it is, at a conservative estimate, five hundred times stronger than cyanide. A single lethal dose of the pure toxin would be about the amount that would rest on the head of a pin.

Tetrodotoxin's role in human history reaches literally to the dawn of civilization. The Egyptians knew of the poison almost five thousand years ago; a figure of a puffer fish appears on the tomb of Ti, one of the pharaohs of the Fifth Dynasty. The deadly Red Sea puffer was the reason for the biblical injunction against eating scaleless fish that appears in the Book of Deuteronomy. In China the toxicity of the fish is acknowledged in the *Pentsao Chin*, the first of the great pharmacopeia, supposedly written during the reign of the mythical Emperor Shun Nung (2838 B.C.–2698 B.C.). In the East there is a continuous record that reflects an increasingly sophisticated knowledge of the biology and toxicology of the fish. By the time of the Han Dynasty (202 B.C.– A.D. 220), it was recognized that the toxin was concentrated in the liver; four hundred years later, during the Sui Dynasty, an accurate account of the toxicity of the liver, eggs, and ovaries appears in a well-known medical treatise. The last of the *Great Herbals*, the *Pentsao Kang Mu* (A.D. 1596), recognizes that toxin levels vary in different species and that within any one species they may fluctuate seasonally. It also offers a succinct but vivid description of the results of eating the liver and eggs: "In the mouth they rot the tongue, if swallowed they rot the gut," a condition that "no remedy can relieve." This is but one of the injunctions mentioned in the herbal warning of the dangers of the fish. Yet the *Pentsao Kang Mu* also reveals an extraordinary development that had taken place in Mandarin society. Despite the obvious risks, by 1596 the fish had become something of a culinary delicacy. Several recipes describe in great detail methods of preparing and cooking the fish that are said to eliminate some of the toxin and render the flesh edible. One account suggested soaking the roe overnight in water; another heralded the delight of eating "salted eggs and marinated testes."

Just how much range of error these methods allowed is uncertain. The herbal also records a folk saying that remains popular to this day in China and Japan: "To throw away life, eat blowfish."

The subtleties of safely preparing puffer fish were quite unknown to the first European explorers to reach the Orient, and as a result they have left some of the most vivid accounts of just what these toxins are capable of. During his second circumnavigational voyage, Captain James Cook ignored a warning from the two naturalists he had on board and ordered the liver and roe of a puffer dressed for his supper. Cook insisted that he had safely eaten the fish elsewhere in the Pacific, and then in the unassuming way of a captain in the Royal Navy, he invited the two naturalists to eat with him. Fortunately, the three men merely tasted the morsel. Nevertheless, between three and four in the morning they were "seized with an extraordinary weakness in all our limbs attended with a numbness or sensation like that caused by exposing one's hands or feet to a fire after having been pinched much by frost." Cook wrote, "I had almost lost the sense of feeling; nor could I distinguish between light and heavy bodies . . . a quart pot full of water and a feather being the same in my hand." Cook and his naturalists were lucky. Two sailors on the Dutch brig *Postilion* rounding the Cape of Good Hope some seventy years later fared less well. This account is offered by the physician who arrived at the bodies not ten minutes after they had eaten the fish (a species of *Diodon*, as it turned out). The boatswain

lay between decks, and could not raise himself without the greatest exertion; his face was somewhat flushed; his eyes glistening, and pupils rather contracted; his mouth was open, and as the muscles of the pharynx were drawn together by cramp, the saliva flowed from it; the lips were tumid and somewhat blue; the forehead covered with perspiration; the pulse quick, small and intermittent. The patient was extremely uneasy and in great distress, but was still conscious. The state of the patient quickly assumed a paralytic form; his eyes became fixed in one direction; his breathing became difficult, and was accompanied with dilation of the nostrils; his face became pale and covered with cold perspiration; his lips livid; his consciousness and pulse failed; his rattling respiration finally ceased. The patient died scarcely 17 minutes after partaking of the liver of the fish.

The sailor's partner suffered the same symptoms, except that he vomited several times, which made him feel momentarily relieved. He expressed some hope until "a single convulsive movement of the arms ensued, whereupon the pulse disappeared and the livid tongue was protruded from between the lips." His death occurred about one minute after that of his shipmate.

While Cook and the rest of the Europeans were having their difficulties on the high seas, the Japanese had adopted the Chinese passion for the puffer fish and carried its preparation to the level of art. The ardor with which the Japanese consumed their *fugu* fish bewildered early European observers. Engelbert Kaempfer, a physician attached to the Dutch embassy in Nagasaki at the turn of the eighteenth century, noted that "the Japanese reckon [this] a very delicate fish, and they are fond of it, but the Head, Guts, bones and all the garbage must be thrown away, and the flesh carefully wash'd and clean'd before it is fit to eat. And yet many people die of it. . . ." He also observed that the fish was so dangerous and yet so popular that the emperor had been obliged to issue a special decree forbidding his soldiers to eat it. Curiously, though Kaempfer seems to have witnessed many individuals eating and enjoying the puffer, he concludes, "the poison of this sort is absolutely mortal, no washing nor cleaning will take it off. It is therefore never asked for, but by those who intend to make away with themselves." This Dutchman, like countless generations of Western visitors who came after him, missed the point of the puffer experience completely. As the Japanese explain in verse, "Those who eat fugu are stupid. But those who don't eat fugu are also stupid."

Today the Japanese passion for puffers is something of a national institution. In Tokyo alone puffers are sold by over eighteen hundred fish dealers. Virtually all the best restaurants offer it, and to retain some semblance of control the government actually licenses the specially trained chefs who alone are permitted to prepare it. Generally the meat is eaten as sashimi. Thus sliced raw, the flesh is relatively safe. So are the testes, except that they are sometimes confused with the deadly ovaries by even the most experienced chefs. Yet many connoisseurs prefer a dish known as *chiri*, partially cooked fillets taken from a kettle containing toxic livers, skins, and intestines. Lovers of chiri are invariably among the hundred or more fatalities that occur each year.

The Japanese prefer and pay premium prices for four species of puffer, all in the genus *Fugu*, and all known to be violently poisonous.

Why would anyone play Russian roulette with such a creature? The answer, of course, is that fugu is one of the few substances that walks the line between food and drug. For the Japanese, consuming fugu is the ultimate aesthetic experience. The refined task of the fugu chef is not to eliminate the toxin, it is to reduce its concentration while assuring that the guest still enjoys the exhilarating physiological aftereffects. These include a mild numbing or tingling of the tongue and lips, sensations of warmth, a flushing of the skin, and a general feeling of euphoria. As in the case of so many stimulants, there are those who can't get enough of a good thing. Though it is expressly prohibited by law, certain chefs prepare for zealous clients a special dish of the particularly toxic livers. The organ is boiled and mashed and boiled again and again until much of the toxin is removed. Unfortunately, many of these chefs succumb to their own cooking. It was such a dish that caused the controversial death in 1975 of Mitsugora Bando VIII, one of Japan's most talented Kabuki actors, indeed, an artist who had been declared a living national treasure by the Japanese government. He, apparently, like all of those who eat the cooked livers, was among those who, in the words of one fugu specialist, enjoy "living dangerously."

Because of its popularity as a food and the relatively high incidence of accidental poisonings, the fugu fish has generated an enormous medical and biomedical literature. Exploring that literature for clinical descriptions and case histories, I was immediately struck by the parallels to the zombi phenomenon. In describing his experience to me Clairvius Narcisse recalled remaining conscious at all times, and although completely immobilized could hear his sister's weeping as he was pronounced dead. Both at and after his burial his overall sensation was that of floating above the grave. He remembered as well that his earliest sign of discomfort before entering the hospital was difficulty in breathing. His sister recalled that his lips had turned blue, or cyanotic. Although he did not know how long he had remained in the grave before the zombi makers came to release him, other informants insist that a zombi may be raised up to seventy-two hours after the burial. The onset of the poison itself was described by several houngan as the feeling in victims "of insects crawling beneath your skin." Another houngan offered a poison that would cause the skin to peel off the victim. Popular accounts of zombis claim that even female zombis speak with deep husky voices, and that all zombis are glassy-eyed. Several houngan suggested that the belly of the victim swells up after he or she has been poisoned.

Once again, recall the list of medical symptoms from the right-hand side of my blackboard. At the time of his reputed death Narcisse suffered digestive troubles with vomiting, pronounced respiratory difficulties, pulmonary edema, uremia, hypothermia, rapid loss of weight, and hypertension. Note that these symptoms are quite specific and certainly peculiar.

Now compare Narcisse's constellation of symptoms with the following specific description of the effects of tetrodotoxin [italics mine]:

The onset and types of symptoms in puffer poisoning vary greatly depending on the person and the amount of poison ingested. However, symptoms of *malaise*, pallor, dizziness, *paresthesias* of the lips and tongue and ataxia develop. The *paresthesias* which the victim usually describes as a *tingling or prickling sensation* may subsequently involve the fingers and toes, then spread to other portions of the extremities and gradually develop into severe numbness. In some cases the numbness may involve the entire body, in which instances the patients have stated that it felt as though *their bodies were floating*. Hypersalivation, profuse sweating, extreme weakness, headache, *subnormal temperatures*, decreased blood pressure, and a rapid weak pulse usually appear early. Gastrointestinal symptoms of nausea, vomiting, diarrhea and epigastric pain are sometimes present. Apparently the pupils are constricted during the initial stage and later become dilated. As the disease progresses the eyes become fixed and the pupillary and corneal reflexes are lost. . . . Shortly after the development of paresthesias, *respiratory distress* becomes very pronounced and . . . the *lips, extremities and body become intensely cyanotic*. Muscular twitching becomes progressively worse and finally terminates in *extensive paralysis*. The first areas to become paralyzed are usually the throat and larynx, resulting in aphonia, *dysphagia*, and *complete aphagia. The muscles of the extremities become completely paralyzed and the patient is unable to move*. As the end approaches the eyes of the *victim become glassy. The victim may become comatose but in most cases retains consciousness, and the mental faculties remain acute until shortly before death*. [See Halstead, Annotated Bibliography.]

Several physicians report this most peculiar state of profound paralysis, during which time most other mental faculties remain normal.

One notes "the patient's comprehension is not impaired even in serious cases. When asked about his experiences he can describe everything in detail after recovery." Other documented and pronounced symptoms of tetrodotoxin poisoning include pulmonary edema, hypotension, cyanosis, hypothermia, nausea and vomiting. Respiratory distress is almost always the first symptom of the poisoning, and many victims develop distended bellies. The third day after exposure to tetrodotoxins, large skin blisters may appear; by the ninth day the skin begins to peel off. A Chinese patient admitted to the Queen's Hospital in Honolulu complained that he felt "numb from neck to toes with a feeling of ants crawling over him and biting him."

This list is not exhaustive. In all, Narcisse shared twenty-one or virtually all the prominent symptoms documented in known cases of tetrodotoxin poisoning.

Not only did the individual symptoms of zombification and tetrodotoxication sound remarkably similar, but entire case histories from the Japanese literature read like accounts of the living dead. A Japanese peddler shared a dish of chiri with several mates and suffered all the classic symptoms of puffer poisoning. The physicians gave up, certain that the man was dead, but the individual recovered, and not fourteen hours after he had eaten the poisonous food he walked out of the hospital. A Korean miner and his son ate the ovaries of a species of *Sphoeroides* and within an hour were taken to the hospital. The father retained "clear consciousness" until he died; his son suffered complete immobility for about two hours but recovered naturally without treatment.

These two accounts illustrate one of the most eerie characteristics of puffer poisoning. Tetrodotoxin induces a state of profound paralysis, marked by complete immobility during which time the border between life and death is not at all certain, even to trained physicians. I need hardly express the significance of this in terms of the zombi investigation. It became quite clear that tetrodotoxin was capable of pharmacologically inducing a physical state that might actually allow an individual to be buried alive.

In Japan, apparently, it had already happened. One physician reported:

A dozen gamblers voraciously consumed fugu at Nakashima-machi of Okayama in Bizen. Three of them suffered from

poisoning; two eventually died. One of these being a native of
the town was buried immediately. The other was from a distant district . . . under the jurisdiction of the Shogun. Therefore the body was kept in storage and watched by a guard
until a government official could examine it. Seven or eight
days later the man became conscious and finally recovered
completely. When asked about his experience, he was able to
recall everything and stated that he feared that he too would
be buried alive when he heard that the other person had been
buried.

What happened to the unfortunate individual who was buried is not
explained. The second case was equally dramatic.

A man from Yamaguchi in Boshy suffered from fugu poisoning at Osaka. It was thought that he was dead and the body
was sent to a crematorium at Sennichi. As the body was being
removed from the cart, the man recovered and walked back
home. As in the case previously cited, he too remembered
everything.

These two cases were by no means unique. In fact such incidences
are apparently frequent enough that in some parts of Japan a person
declared dead from eating puffer fish is customarily allowed to lie
alongside his or her coffin for three days before burial. On Christmas
Eve 1977 a forty-year-old resident of Kyoto was admitted to a hospital
after being poisoned by fugu. The patient soon stopped breathing, and
all symptoms were consistent with brain death. Physicians immediately
initiated artificial respiration and other appropriate treatments. These
did not help, but twenty-four hours later the patient spontaneously
began to breathe. He eventually recovered completely, and later remembered hearing his family weeping over his still body. His senses
were unimpaired. He wanted desperately to let them know that he was
alive, but he was unable to. "That," he later told medical investigators,
"was really hell-on-earth."

These reports cast the zombi investigation in a totally different
light. Suddenly it seemed not only possible but likely that Marcel's
poison could cause a state of apparent death. Now a dozen more specific questions came to mind. Did the species of puffers used by Marcel
contain the critical toxin? If so, could they have survived the preparation? Recall that Marcel had placed the dried fish on a charcoal grill

and broiled them to an oily consistency; heat destroys many chemical compounds. What about the way the zombi poison was reputedly applied? How would the bokor assure that the victim did not die from the poison? Once again, many of these questions could be answered from the literature.

Marcel added two species of puffers to his poison—*Diodon hystrix* and *Sphoeroides testudineus*—both known to contain tetrodotoxin. It was a species of *Diodon* that poisoned the Dutch sailors on the *Postilion* as it rounded the Cape of Good Hope. Members of the genus *Sphoeroides* are closely related to the Japanese fugu fish, and are known to be particularly virulent. In the mid-1950s an elderly tourist in south Florida ate the liver of *Sphoeroides testudineus*. Forty-five minutes later she died, after suffering all the horrible symptoms of the disease. Clearly, the species Marcel had used could contain tetrodotoxin.

Other answers were suggested by a remarkable account handed down by the Mexican historian Francisco Javier Clavijero. In 1706, while searching for a new mission site in Baja California, four Spanish soldiers came upon a campfire where indigenous fishermen had left a roasted piece of the liver of a botete (*Sphoeroides lobatus*). Despite the warnings of their guides, the soldiers divided the meat. One of them ate a small piece, another chewed his portion without swallowing, and the third only touched it. The first died within thirty minutes, the second shortly thereafter, and the third remained unconscious until the next day.

Two critical points came across. That the soldiers were poisoned by roasted meat exemplifies the important fact that heat—frying, boiling, baking or stewing—does not denature tetrodotoxins. Secondly, although tetrodotoxin is one of the most poisonous chemicals known, like any drug its effects depend on dosage and the way it is administered. Having studied over a hundred cases of tetrodotoxication, the Japanese investigators Fukada and Tani distinguished four degrees of poisoning. The first two they characterized by progressive numbing sensations and the loss of motor control; the equivalent, perhaps, of having one's entire body "fall asleep." The third degree includes paralysis of the entire body, difficulty in breathing, cyanosis, and low blood pressure—all suffered while the victim retains clear consciousness. In the final degree, death comes very quickly, as a result of complete respiratory failure. If the poisonous material is ingested, the onset of the third degree is usually very rapid. The sailors on the *Postilion* typically died within seventeen minutes. If tetrodotoxin somehow enters the bloodstream directly its potency is enhanced forty to fifty times. However, tetrodotoxin is also topically active, and some of the preliminary symptoms have shown up in individuals who merely handle the toxic organs. In other words, whether or not the victim of the zombi poison survived would depend on just how he or she was exposed to the poison.

Marcel had recognized the potency of his preparation and had acknowledged at least implicitly the importance of proper application and correct dosage. He most emphatically stated that it was never placed in the victim's food. Now I understood why. Ingested orally it would more than likely kill the victim; applied repeatedly to the skin or open wounds, or blown across the face of the victim so that it was inhaled, it could bring on a state of apparent death.

One final point was critical. Those who are poisoned by tetrodotoxin generally reach a crisis after no more than six hours. If the victim survives that period, he or she may expect a full recovery, at least from the effects of tetrodotoxin. This made it at least theoretically possible for a poison victim to appear dead, be hastily buried, and then recover in the coffin.

The implications of these conclusions were extraordinary. Here was a material basis for the entire zombi phenomenon—a folk poison containing known toxins fully capable of pharmacologically inducing a state of apparent death. That the peculiar symptoms described by Clairvius Narcisse so closely matched the quite particular symptoms of tetrodotoxin poisoning suggested that he had been exposed to the poison. If this did not prove that he had been a zombi, it did, at least, substantiate his case. And there was one more especially haunting fact. Every indication pointed to the possibility that Narcisse had remained conscious the entire time. Totally paralyzed, he may have been a passive observer of his own funeral.

As soon as I had these results I contacted Nathan Kline and, on his instructions, forwarded an unmarked sample of the poison to a Professor Leon Roizin at the New York State Psychiatric Institute. The next step was straightforward. Before initiating an expensive series of chemical studies, Roizin would see what the powder could do to laboratory animals. To avoid any possible bias in the experiments, I told Roizin neither what the powder contained nor what it was reputedly used for. He was merely instructed to prepare an emulsion and apply it topically to the animals. I heard from him within a week. He had worked very fast, and was ready to see me.

The elevator opened onto a congested hall that smelled of laboratories. I took a few steps along the short corridor and paused, flanked on both sides by photographic montages of monkeys under the influence of megadoses of various drugs. I was in no position to cast judgment, but I had to acknowledge that these photographs were more gruesome than anything I had witnessed in any bokor's collections.

Roizin worked in a cluttered office. He was short, almost dwarfed by his white lab coat. Discussing his experiments, we spoke clinically.

"What you have here," he began, "is a very strong neurotropic substance. Most peculiar. You have no idea what it contains?"

I shook my head. I didn't like to deceive him, but I wanted first to hear what he had discovered. Explanations could come later.

Roizin reached into a folder and passed me two photographs of white rats. Both appeared dead. "First we shaved the dorsal surface and applied a thick emulsion containing the powder. Within a quarter hour spontaneous activity decreased, and within forty minutes the animals moved only when stimulated. Soon after that all mobility ceased, and the rats held a single position for three to six hours. Breathing became superficial. We could still get a heartbeat, and there was some response to sound and corneal stimulation. After six hours all movement completely ceased. They looked comatose and showed no response to any stimuli. Yet we could still register a heartbeat on the EKG. We also picked up brainwaves. They remained that way for twenty-four hours."

"And they were alive?" I asked.

"Yes, but totally tranquilized. But there is much more. Most tranquilizers do not act as anesthetics. This one does. Look at the second photograph." I glanced down. "We stuck that needle into its tail, and there was absolutely no response of any kind. Whatever you have here acts peripherally as an anesthetic. And it acts very quickly, without affecting the heart even in the toxic stage."

"Were there any other behavioral changes?"

"Yes, one that comes to mind. Early in the experiment while there was still some mobility, the rats clustered together, and their extremities were noticeably cool. It was as if they were attempting to keep warm."

"How much of the powder was administered to each animal?"

"Five milligrams per one hundred gram weight. But there is something else I want to show you." Roizin lifted his phone to make a call. I took the opportunity to discreetly make a calculation on my notepad based on the dosage administered to rats. Extrapolating to the weight of an average man, Roizin had used the equivalent of 3.5 grams of the crude poison.

"The technician is ready down in the studio," he said as he hung up. "When we got these results with the rats I decided to record what happened to the monkeys." He got up and indicated I should follow him out of his office. "Ever worked with rhesus monkeys before?" he asked as he led me away.

"Never had to," I replied.

"You're most fortunate, then. But you ought to experience the animals at least once before you see the film." Roizin took me through a maze of cinderblock corridors that led to the animal facility of the institute. The sound as we approached the room that housed the cages of monkeys was deafening. The last time I had heard any monkey had been in the rain forest of the northwest Amazon. But that had been the deep roar of a howler, free and majestic and fearsome. Here, an entire wall of metal cages rattled with the frantic movements of animals that had never called a mate. Out of the dozen sterile screams, Roizin picked one and tapped its cage menacingly. The captive lunged violently, knocking its teeth against the thin bars. "Passive they are not," he said loudly and close to my ear. His point made, he turned out of the room.

"What you will see, of course, are but preliminary results," he told me as we entered the studio. "We used exactly the same procedure as with the rats." A technician dimmed the lights, and the fuzzy image of Roizin came on the video monitor. Beside him was a single cage, with an aggressive rhesus monkey just like those we had seen. Roizin's voice on tape was difficult to understand, but the images spoke for themselves. Twenty minutes after application of the emulsion this same monkey appeared noticeably sedated. Even when prodded it did not lunge at the bars. Rather, it slowly retreated to one corner of the cage, assuming, as Roizin explained, a stereotypic catatonic position. When stimulated, it reacted at best passively, opening its mouth in a vain imitation of its normal threatening posture. The monkey maintained that position for three hours. At that point the technician and Roizin had been called away on other business. When they returned six hours later the monkey was still in the same position.

"Now we can increase dosage and do any number of things," Roizin told me once we were back in his office, "but I think for now the point is made. Whatever this powder contains, it acts very quickly and completely modifies behavior." Roizin leaned back in his chair. "I can promise you one thing for certain. For many years I have worked for Professor Kline, and some pretty odd drugs have passed through this laboratory. This, without doubt, is the most peculiar."

"Do you have any idea what it might be used for?" I asked.

"There's no way of knowing right now."

"Don't you have any guesses?"

"Perhaps cardiovascular surgery. It is curious how the heart remains unaffected while the body is totally anesthetized. Also in psychiatry, it might be of use treating something like psychotic excitement."

"As a tranquilizer?"

"Of a sort. There is one more thing. Did Nate Kline ever talk to you about experimental hibernation?"

These preliminary lab results should have delighted me. They demonstrated experimentally what all the literature could only suggest as a possibility. Marcel's powder did contain pharmacologically active compounds that acted very rapidly to lower the metabolic rate of the victim. Yet even as I received congratulatory letters and calls from Kline and Lehman, and from Schultes as well, I was more deeply perplexed than ever. Before these results the entire notion of zombis had remained strictly an idea, a curiosity, an abstraction. I went to Haiti skeptically, knowing nothing about the country or the people, and the assignment dropped me into an enchanted land whose spiritual rhythms took me utterly by surprise and moved me profoundly. Despite this, or perhaps because of it, I had never actually paused to consider whether or not zombis truly existed. It wasn't that I didn't believe, and it wasn't that I did. I just hadn't passed judgment. The formula of the poison, the correlations from the literature and the case of Narcisse, and now the preliminary but concrete laboratory results changed everything. Now I had to face just how little I understood about a phenomenon that suddenly appeared hauntingly real.

There were so many loose ends. Every report received from Haiti, for example, had mentioned an antidote. Yet while Marcel had prepared a potion that he believed counteracted the effects of the poison, the way it was used and the ingredients themselves suggested that it was pharmacologically inactive. Tetrodotoxin has no known medical antidote, nor, from what is known of the action of the toxin, would the zombi makers require one. Victims of puffer fish poisoning either live or they die, and those who survive recover on their own, as presumably would the zombis. Still, this information notwithstanding, the persistent reports of an antidote begged further investigation. As colleagues often remind me, absence of evidence is not evidence of absence.

Yet in all this obsession with the drug and elusive antidote, I was, in one sense, missing the point entirely. All that the formula of the poison explained was how an individual might be made to appear dead. Clearly the same thing occurred in Japan, however infrequently, but just as clearly those who succumbed to toxic fugu preparations were not zombis; they were merely poison victims. Any psychoactive drug has within it a completely ambivalent potential. Pharmacologically it induces a certain condition, but that condition is mere raw material to be worked by particular cultural or psychological forces and expectations. This is what experts call the "set and setting" of any drug experience. *Set* in these terms is the individual's expectations of what the drug will do to him; *setting* is the environment—both physical and, in this case, social—in which the drug is taken. For example, in the northwest rain forests of Oregon there are a number of native species of hallucinogenic mushrooms. Those who go out into the forest deliberately intending to ingest these mushrooms generally experience a pleasant intoxication. Those who inadvertently consume them while foraging for edible mushrooms invariably end up in the poison unit of the nearest hospital. The mushroom itself has not changed.

This did not suggest that the zombi poison might be only a pleasant hallucinogen. But like the mushroom, its potential was latent. The Japanese victim lying conscious but paralyzed while his family mourned his death might, upon recovery, rationalize his terrifying experience within the expectations of his society. Everyone knows that is what fugu poisoning is like. Without doubt, in the phantasmagoric cultural landscape of Haiti, Clairvius Narcisse had his own expectations that he carried with him literally into and out of the grave. Of what was going on in his mind, I had no idea, and until I did I would know nothing about zombis. But I did have available to me—and the opportunity to explore it before my return to Haiti—a literature that would provide a context for understanding.

8

Voodoo Death

COUNT KARNICE-KARNICKI was a compassionate man, and his invention made him the rage of Europe. The count was a Russian nobleman, the chamberlain to the czar, but his inspiration had come in Belgium while attending the funeral of a young girl. As the first shovelfuls of dirt landed on the wooden coffin, a pitiful scream rose from the earth, staggering the officiating priest and causing a number of young women to faint. It was a sound that the count would never forget. He, like so many of his generation in all corners of Victorian Europe, became obsessed by the threat of premature burial.

His invention, introduced just before the turn of the century, was a simple contraption, efficient and inexpensive enough to be well within reach of rich and poor alike. For the truly destitute, the apparatus was available for rent. It consisted of a hermetically sealed box and a long tube that would be fixed into an aperture in the coffin as soon as it was lowered into the ground. On the chest of the dead person was placed a large glass ball attached to a spring linked to the sealed box. With the slightest movement of the glass ball, as would occur if breathing began, the spring would be released, causing the lid of the box to fly open and admit both light and air to the buried coffin. At the same time the spring initiated a mechanical chain reaction worthy of Rube Goldberg. A flag sprang four feet above the box, a bell began to ring and continued for thirty minutes, and an electric lamp ignited. The long tube was envisioned not only to admit oxygen, but also to serve as a megaphone, amplifying the presumably weak voice of the almost dead. Not a hundred years ago, at the turn of this century, this peculiar apparatus was heralded as a technological breakthrough. Many thousands of Frenchmen left specific instructions in their wills to ensure that it would be placed on their tombs. In the United States it was so popular that societies formed to promote its subsidized use.

The citizens who embraced Count Karnice-Karnicki's invention were responding to an epidemic of premature burials that had colored the popular press and confounded medical authorities. A typical report appeared in the *London Echo* in March of 1896. Nicephorus Glycas, the Greek Orthodox metropolitan of Lesbos, was pronounced dead in his eightieth year. According to the traditions of his church he was immediately garbed in his episcopal vestments and placed on a throne, where his body was exposed day and night to the faithful, and guarded constantly by priests. On the second night, the old man suddenly awoke and stared with amazement and horror at the parade of mourners at his feet. His priests, according to the report, were no less startled to realize that their leader had not been dead but had merely fallen into a deathlike trance. The *Echo* correspondent impassionately questioned what might have happened had the metropolitan been a layman, and then concluded that he would have been buried alive.

A second popular account was that of a Reverend Schwartz, an Oriental missionary who was reportedly aroused from apparent death by his favorite hymn. The congregation celebrating his last rites was stunned to hear a voice from the coffin joining in on the refrain.

Though today both of these cases may appear preposterous, at the time they were not only seriously discussed, they were believed, and they helped fuel a hysterical fear of premature burial that swept late Victorian Europe. In 1905 an English physician and member of the Royal College of Surgeons edited a volume in which were documented 219 narrow escapes from premature burial, as well as 149 cases in which the body was actually interred while still alive. Also noted were ten instances in which autopsies were erroneously performed on the living, and two cases in which consciousness returned to the "corpse" during the process of embalming.

Many people were not about to take any chances. Hans Christian Andersen constantly carried a note in his pocket instructing what should be done with his body in the event of his death. The English novelist Wilkie Collins placed a similar precautionary note by his bedside table each night. So did Dostoyevsky, who urged that his burial be delayed five days lest his apparent death be but a trance. Certain leading members of the British aristocracy took more drastic measures, which incidentally were remarkably similar to certain Haitian practices that their countrymen would no doubt have condemned. The vodounist, fearing that a family member will be raised as a zombi, has been reported to drive a blade through the heart of the dead. Fearing premature burial, the noted British antiquary Francis Douce requested in his will that the surgeon Sir Anthony Carlisle be permitted to sever his head from his body. So did a certain Harriet Martineau. A well-known actress of the era, Ada Cavendish, left instructions in her will that her jugular be sliced. Lady Burton, widow of the famous African explorer and writer Sir Richard Burton, provided that her heart be pierced with a needle. Bishop Berkeley, Daniel O'Connell, and Lord Lytton had similar fears and ordered that their burials be delayed, and that one or more of their veins be opened so that their blood would drain and thus assure that they were truly dead.

By the turn of this century fear of premature burial had grown into an overriding public concern. It was discussed in all the learned medical journals, and in England it generated parliamentary inquiries that led to the Burial Act of 1900, which among its many statutes specified the length of time that had to transpire between the pronouncement of death and actual interment. On the Continent prizes were offered for the discovery of a conclusive sign of death. In France in 1890 a certain Dr. Maze was awarded the prestigious Prix Dusgate and twenty-five hundred francs simply for asserting that the only reliable sign of death was putrefaction. The earnest scientific interest in establishing the difference between real and apparent death is evident in an academic textbook on the subject published in 1890 whose bibliography lists no fewer than 418 citations.

Actually, a satisfactory means of diagnosing death has both obsessed and eluded man since earliest times. Of course the fundamental signs of death have always been known, and they have not changed.

They are: cessation of respiration and heartbeat, changes in the eye, insensibility, rigor mortis, pallor and discoloration due to the settling of the blood. The problem, as Kline had expressed so forcefully when we first met, has always been that not one of these is foolproof. And once that is recognized and admitted, a floodgate of possibilities opens.

But there was something else going on in the minds of Victorian society. Just how serious a threat premature burial was in the late nineteenth century is uncertain. Even at the time many insisted that reports were greatly exaggerated. Yet the very fact that the debate flourished in the Houses of Parliament and the halls of the Royal Academy was as significant as its outcome. An already uneasy public undoubtedly took note that the subject was being seriously considered within institutions that were the very pillars of the Victorian world and, by extension, of reason itself. As physicians outlined the difficulties of diagnosing death, politicians debated how long the dead should be kept from the grave, and salesmen pitched Count Karnice-Karnicki's invention, the public mood was further aroused by certain popular accounts. One of these was the notorious case of a Colonel Townsend. According to a panel of physicians called upon to witness the event, this officer willingly reduced his heart rate and entered a self-induced trance, or, as some described it, a state of suspended animation. Heartbeat ceased, respiration stopped, and the entire body assumed the icy chill and rigidity of death. The color fled from Townsend's face; his eyes became glazed and fixed. After he had been comatose for thirty minutes, the physicians actually certified him dead and prepared to go home. As they did so, Townsend began to recover slowly, and by the next day he was well enough to repeat his feat. This case was widely quoted not only in the press but in academic textbooks of medical jurisprudence, and it undoubtedly lent credence to contemporary statements such as "the difference between trance and death has never been quite understood by the majority of mankind."

This statement perfectly encapsulated the Victorians' dilemma. At the root of the hysterical fear of premature burial was the fact that physicians recognized, and patients suffered, a number of peculiar conditions characterized by immobility and insensibility, and known variously as trance, catalepsy, cataplexy, and suspended animation. As far as the public was concerned, any one of these clinical diagnoses could be the ominous prelude to accidental interment. Victorian physicians noted that catalepsy was marked by the singular absence of will or volition; the body of the patient remained in whatever position it was placed. Trance was said to most nearly resemble the condition of a hibernating animal, with the patient suffering complete mental inertia. Cataplexy was a modification of the same conditions, only the patient simply collapsed limply to the ground with the eyes closed, completely immobile, unable to speak, yet conscious and totally aware of all that was going on around him. Medical writers of the era even discussed a fourth clinical condition, marked by immobility, which they termed ecstasy, but this was not a state that might lead to premature burial. The ecstatic patient was described typically as having a "radiant, visionary expression and a tendency to fix himself in statuesque poses whilst concentrating upon some object of adoration."

Needless to say, these conditions are no longer recognized by the medical profession. Certain aspects of the cataleptic condition have been subsumed under "catatonic schizophrenia," but trance has been reduced to a feature of hypnosis research, and ecstasy and cataplexy have disappeared as clinical diagnoses. But for the Victorians these ailments did exist, and they were discussed seriously by the leading medical authorities precisely because people were succumbing to them. Where did they come from? Cataplexy, for example, is described in the old medical textbooks as "being precipitated by strong emotion and persisting until such emotion be controlled," which sounds not unlike another common feature of Victorian life, also now abandoned—the fainting spell. Recently some have suggested that women of the era simply suffered the physical consequences of wearing impossibly tight corsets, but this interpretation misses the point. Fainting was a socially conditioned response; in certain clearly recognized and predictable situations it was virtually expected. Young women of the elite, in particular, found fainting a convenient means to avoid or modify uncomfortable social predicaments. Some learned to attain their desires simply by cleverly faking, while others actually did pass out, and in some cases it was assumed by physicians that they were dead. In other words, a socially conditioned act became a physiological reality.

Like the fainting spell, catalepsy, cataplexy, trance, and ecstasy were socially conditioned ailments, and their cause lay somewhere deep within the psyche of the age. Their manifestations were concrete and isolated to a particular time. The fear of premature burial undoubtedly was accentuated because people really believed such states of sham death possible. And just because they did exist for the Victorians, some unfortunate souls may well have ended up in the ground wishing that their relatives had rented from Count Karnice-Karnicki.

Part of what was going on in Victorian England was related to a phenomenon that Western anthropologists had noted in "primitive" societies but overlooked in their own culture. For just as an individual's sickness may have a psychosomatic basis, it is possible for a society to generate physical ailments and conditions that have meaning only in the minds of its people. In Australia, for example, aborigine sorcerers carry bones extracted from the flesh of giant lizards, and when these slivers are pointed at a person while a death spell is recited, the individual invariably sickens and almost always dies. According to one scientific report the victim

> stands aghast, with his eyes staring at the treacherous pointer, and with his hands lifted as though to ward off the lethal medium which he imagines is pouring into his body. His cheeks blanch and his eyes become glassy and the expression on his face becomes horribly distorted . . . he attempts to shriek but usually the sound chokes in his throat, and all that one might see is froth at his mouth. His body begins to tremble . . . he sways backwards and falls to the ground . . . writhing as if in mortal agony. After awhile he becomes very composed and crawls to his [shelter]. From this time onwards he sickens and frets, refusing to eat and keeping aloof from the daily affairs of the tribe.

At this point only the nangarri, or medicine man, may save him by initiating a complex ritual. But should the nangarri refuse to cooperate, the victim will almost certainly die.

What happens to the Australian aborigine is an example of something that occurs in many cultures. It is a phenomenon every bit as real, and every bit as enigmatic, as the ailments generated by the Victorian mind. Its basic pattern is consistent. An individual breaks a social or spiritual code, violates a taboo, or for one reason or another believes himself a victim of putative sorcery. Conditioned since childhood to expect disaster, he then acts out what amounts to a self-fulfilling prophecy. Often the death knell is sounded by a hex or, as in Australia, a simple gesture rife with meaning. Sorcerers may use props as media of transmission: African witch doctors have knucklebones, and European witches carved wooden dolls. Or transmission is direct. Even today in Greece the harbinger of death need merely squint the evil eye.

In Haiti, there are literally dozens of methods, and that is perhaps why anthropologists call the entire phenomenon voodoo death.

Voodoo death has been so commonly reported, and so frequently documented and verified by scientific observers, that its existence is no longer a matter of debate. Of course, no scientist would believe that there is a direct causal relationship between the death of the victim and, for example, the physical act of pointing a bone. Clearly it is the victim's mind that mediates the sorcerer's curse and the fatal outcome. What remains to be discovered is the mechanism that actually allows this to occur. Three possible explanations have been offered.

The first scientists seriously to consider the phenomenon of voodoo death were not anthropologists but physicians, many of them affected by the peculiar cases they witnessed on the battlefields of Europe during the First World War. In the nightmare of despair and death on the Western Front, certain traumatized soldiers who had not suffered any wound inexplicably died of shock, a medical disorder normally brought about by a critical drop in blood pressure due to excessive bleeding. When these physicians later became familiar with cases of voodoo death, they saw a connection. They suggested that individuals terrified by a magic spell suffered, like the soldiers, from an overstimulation of the sympathetic-adrenal system, which led to a form of fatal shock. Fear, in other words, could initiate actual physiological changes that quite literally led to death.

Many anthropologists, less familiar with the complex workings of the autonomic nervous system, have considered voodoo death as a psychological process, emphasizing the power of suggestion. If faith can heal, they argue, fear can kill. Psychologists have studied, for example, something that most of us take for granted—that the likelihood of becoming ill or even dying depends to a large extent on our frame of mind. Feelings of depression, hopelessness, or despair do not cause diseases, but somehow they make us vulnerable. Loneliness would seem hardly a fatal affliction, yet a disproportionate number of spouses die in the first year after the death of their mates. Psychologists label this the "giving up/given up complex." According to this view, the victim of voodoo death becomes caught in a vicious cycle of belief that indirectly kills him, perhaps, as some suggest, by making his body susceptible to pathogenic disease. His psychological state can be imagined. He is doomed to die by a malevolent curse that both he and all those around him deeply believe in. He becomes despondent, anxious, and fearful. His resignation is both recognized and expected by other members of his society. They join him in speculating how long he may survive, or who is the source of the curse. And then a strange thing happens. A consensus is reached that the end is near, and his friends and family retreat as from the smell of death. They return, but only to wail and chant over the body of this person they consider already dead. Physically the victim still lives; psychologically he is dying; socially he is already dead.

A third group, also anthropologists, agree with this perspective but carry it further by suggesting an actual mechanism to account for physical death. They point out that in many cases the victim of voodoo death is not merely a benign presence but, having by definition crossed into the realm of the spirits, has become an actual threat that must be removed. And in the case of the Australian aborigines, this is precisely what happens. Weakened by the long ordeal, the victim of sorcery receives no relief from even his closest relatives. On the contrary, these former supporters actually take food and water away, on the theory that a dead person has no need for either and with the motive, as one physician was told, "if real close up finish, take water away so spirit goes." In the deserts of Australia, where the daytime temperatures average over one hundred degrees Fahrenheit in the shade, death by dehydration occurs in about twenty-four hours.

Not all cases of voodoo death are as clearly explicable as these examples reported from Australia, where because of the harsh climate a relatively simple act by the kin eliminates the victim's life supports. More usual in voodoo death, but again more enigmatic, the victim dies *despite* the fact that his family offers succor.

All that can be ascertained is that voodoo death occurs, and that as a process it involves a number of complementary factors. Fear probably does initiate physiological changes. Certainly it makes the victim psychologically vulnerable, and this in turn affects the physical health. Neurophysiologists still do not fully understand the process, though the response of the victim's family and society would seem inevitably to influence both his psychological and his physical well-being. So, while a universal mechanism to account for voodoo death has not been identified, the basic assumption is clear. As one researcher has put it, the brain has the power to kill or maim the body that bears it.

●

The metamorphosis of Clairvius Narcisse from human to zombi was a very special instance of voodoo death. A sorcerer's spell initiated a long process that exploited the victim's greatest fears, mobilized the reinforcing beliefs of the community, and finally led to actual death. To the Haitian peasants Narcisse really did die, and what was magically taken from the ground was no longer a human being. Like many sorcerers around the world, the bokor that spun his death had a prop— in this case an ingenious poison that served as a template upon which the victim's worst fears might be amplified ten thousand times. Still, in the end, it was not the powder that sealed Narcisse's fate, it was his own mind.

Consider for a moment what he went through. As a Haitian peasant he had been socialized since childhood to believe in the reality of the living dead. This conviction had been enforced throughout his life by both a complex body of folklore and, more importantly, the direct testimony of friends and family; in Haiti virtually everyone has a vivid zombi tale to tell. For Narcisse, a zombi was a being without will, on the very frontier of the natural world, an entity that could manifest itself as either spirit or human. Zombis do not speak, cannot fend for themselves, do not even know their names. Their fate is enslavement. Yet given the availability of cheap labor, there would seem to be no economic incentive to create a force of indentured service. Rather, given the colonial history, the concept of enslavement implies that the peasant fears and the zombi suffers a fate that is literally worse than death—the loss of physical liberty that is slavery, and the sacrifice of personal autonomy implied by the loss of identity. Critically, for Narcisse as for all Haitian peasants, the fear is not of being harmed by zombis, but rather of becoming one. And it is to prevent such a horrid fate that the relatives of the dead may reluctantly mutilate the corpse if there is any suspicion of foul play. Unless, of course, the family itself was involved in the zombification.

Not only did Narcisse believe in zombis, he undoubtedly was aware of how and why they were created. When his world began to close in on him, he was already personally isolated. Within his lakou he had been ostracized because of his antisocial behavior; within his own family he was actively engaged in a dispute with his brother over the question of the right to sell heritable land. Eventually it was, by all accounts, that very brother who sold him to the bokor. Had Narcisse been in the right, and was zombified by his adversary without the support of the community, it is difficult to imagine that the brother would have been tolerated in the community for close to twenty years. But in fact he was, and even today Narcisse is not. In all likelihood, at the time of his demise, Narcisse had support from neither his immediate society nor his kin; his closest relatives may have been his greatest enemies. And since family members were involved, gossip and rumors undoubtedly took their toll, especially when he began to suffer physical symptoms that he had never known. Gradually as these symptoms got worse and worse, he would have realized that he had become a victim of sorcery. And more than likely he knew why.

His symptoms were real and concrete, and they worsened. He consulted houngan, but they did nothing. By now desperate, he entered the alien environment of the Schweitzer Hospital, knowing something the doctors did not. His condition deteriorated rapidly, and then something more extraordinary occurred. We have noted the devastating impact that social death has on victims of voodoo death throughout the world. The Haitian model takes this one step further. Narcisse was actually pronounced dead in a hospital by Western-trained physicians. More incredibly, from the known actions of the toxins in the poison, and from his own testimony, there is every reason to believe that he remained conscious for much of the time. He actually heard himself pronounced dead, was aware of his sister's weeping, saw the sheet lifted over his face. Like the Japanese victims of tetrodotoxin, he strove desperately to communicate, but with the paralytic poison, he found it impossible.

Then Narcisse entered another realm. Having caused such a dramatic, virtually complete, reduction in metabolic rate, the poison took its victim quite literally to the frontier of death. Indeed, it very nearly killed him . . . as it may have killed many others. His symptoms remained consistent, yet at one point there seems to have been a qualitative change. Perhaps not surprisingly, the advanced symptoms of known tetrodotoxin poisoning merge with those of what Western physicians have termed the autoscopic near-death experience (NDE). Recall once again his description. He sensed that he was floating above his body at all times. When they placed him in the cemetery, he remained above his tomb, still floating, constantly aware of everything that was going on. He was content, he was without fear. He sensed that his soul was about to take a great journey, and it did travel, he insisted, taking in great passages over the land, timeless passages, immaterial yet powerfully real. His travels were multidimensional, yet they always returned him to the gravesite. His notion of time was lost. His tomb was the only axis of his existence.

Strange things happen to us when we die, at least if we are to believe the word of those who come back. Those who have been close to death speak of an ineffable dimension where all intuitive sense of time is lost. Like a dreamscape, it is timeless, but unlike a dream it is impossibly real, a place of crystal awareness wherein the process of death is acknowledged as something positive, calm, even beautiful.

Like Narcisse, virtually every medical patient who has been to the frontier of death experiences a profound separation between his material body and an invisible, nonmaterial aspect of himself, one that often hovers above the flesh; and in nearly all cases the patient identifies not with the body, but with the spirit. An elderly woman who nearly died during severe complications following surgery in a Chicago hospital wrote her physician, "I was light, airy, and felt transparent." Often patients distinctly remember floating above their bodies, looking down at their material selves. A cardiac patient noted, "I was going up slowly, like floating . . . I was looking from up, down . . . they were working the hell out of me." Also typical is the amazement of a construction worker from Georgia following cardiac arrest: "I recognized *me* lying there . . . [it was] like looking at a dead worm or something. I didn't have any desire to go back to it." Sometimes survivors of autoscopic near-death experiences recall conversations between attending physicians and nurses. Often they describe their frustration at not being able to communicate with others physically present at their bedside. "I tried to say something," one patient remembered, "but she [the nurse] didn't say nothing . . . she was like looking at a movie screen that can't talk back and that doesn't recognize you're there. I was the real one and she was unreal. That's the way I felt." Certain survivors describe an extraordinary ability to "travel" through space and time. "It was just a thought process," one explains. "I felt like I could have thought myself anywhere I wanted to be instantly. I could do what I wanted to . . . it's realer than here, really."

And then, those who go through a near-death experience and survive share one thing in common: they all retain a distinct awareness that at one point their immaterial aspect returned to its physical body. It was at this moment, many remember, that they regained consciousness. In hospital patients, this often occurs instantaneously, coinciding with a particular resuscitative procedure. A cardiac patient responded to electric shock: "She [the nurse] picked up them shocker things . . . I seen my body flop like that . . . it seemed like I was up here and it grabbed me and my body, and forced it back, pushed it back." Another experience ended with the sudden arrival of a loved one. The patient explained, "I was up at the ceiling. . . . Then when someone in the family came to the door and called . . . I was instantaneously back in my body."

This, of course, is what Clairvius Narcisse remembers happening. One moment he was floating above his tomb, and then he heard someone call out. But for Narcisse, the voice did not come from a loved one, and when he returned to his body it was not in a hospital bed, it was in a coffin. And for him, the ordeal was only about to begin.

PART THREE

The Secret Societies

9

In Summer
the Pilgrims Walk

"EVEN AS YOU SEE ME, I passed beneath the earth," Marcel confided as we shared another drink in his small bedroom at the Eagle Bar. "It was the same powder that I gave you. It can cause one hundred and one ailments. Mine was a disease of heat. I sweated even in the ocean. It cooked my blood until my veins ran dry, and then it stole the breath from my lungs." He was describing for me his own exposure to the poison.

There was some kind of disturbance in one of the rooms, and Marcel got up from the bed and slipped back into the bar. His place looked good. There was a new sign out front and a fresh coat of paint on most of the rooms. Rachel reached forward to fill my glass.

"Some party," I said. The rum warmed my tongue and the back of my throat. Both of us were still damp with sweat. Marcel had been happy that we had come to see him as soon as I returned to Haiti, and when he'd found out that the powder had worked his excitement had spilled over into a celebration. One of his women had unlocked the Wurlitzer, and the place had become pretty wild. Marcel and Rachel had danced a mad salsa while I'd been swung around by the women and some of the men, and with the pounding jukebox and the sweltering summer heat, sweat had soon greased the concrete floor. Now things had settled down, but there was still a licentious air to the place, stronger than usual.

"Is it strange to be back?" Rachel asked me.

"I don't think so."

"What do you mean?"

"It's difficult to say. Sometimes when you travel a lot, the landscapes pile up so fast that you lose all sense of place."

"And faces?"

"Yes."

Rachel started fooling with the heel of one of the cheap shoes she always wore. "You know they say Marcel paid fifteen thousand dollars to get cured," she said.

"Where'd he get that kind of money?"

"There used to be money. He worked the docks. Some kind of racket with the tourist boats. Marcel was a big Ton Ton and he had some position in the port."

"Oh," I said.

"Then he went too far. Too many beatings, too many in jail."

"Where'd you pick up all this?"

"My uncle, mostly. He was the prefect."

"I see."

"Nobody thought that he'd live. They passed the powder, and for three days they thought he was dead. There was even a wake. His face was all bloated, and his belly swelled up, and my uncle said that if you pricked his skin, you didn't find blood, you got water."

Marcel returned bearing a noxious-looking bottle containing a thick, viscid liquid. "You see," he said, "it still lives. This was my blood. When they sent the powder all the animals came into me and laid eggs. A cockroach came from my nose, and I passed two lizards from behind."

"There was nothing you could do?"

"No. If your force is strong you can resist a coup l'aire, but the powder is different. If you're wrong it gets you, if you're right it gets you."

"But how can they direct it at you alone?"

"The bones. It's only the soul from the graveyard that has that power. That's why the bones are so dangerous."

"But you survived."

"I passed through the hands of thirteen houngan. They drained all the bad blood from my foot. I am still not well. Look at my face. I live only because of the strength of the spirit that calls for me."

Marcel recounted how there had been many treatments, but only the last had saved him. It had been administered by a mambo, a priestess in the Artibonite. On the critical night she bound his jaw, placed cotton in the nostrils, and dressed him in the clothing of the dead. His feet were tied, as were his hands, and he was laid in a narrow trough dug into the ground of the hounfour. A white sheet covered his entire body. A pierre tonnerre and the skulls of a human and a dog were placed on top of the sheet, the sucker of a banana plant beside him. Seven candles cradled in orange peels surrounded the "grave," and calabashes rested by his head, on his abdomen, and at his feet. These three offerings, Marcel suggested, represented the sacred concepts of the crossroad, the cemetery, and Grans Bwa, the spirit of the forest. The mambo straddled the grave, and in a high-pitched wail she invoked Guede, the spirit of the dead. She took a living chicken and passed it slowly over his body, then broke each of the bird's limbs to extract the death spirit from the corresponding limbs of the patient. Next she took the head of the chicken in her mouth and bit it off that she might touch Marcel with its blood. Marcel then felt the contents of the calabashes massaged into his skin, water splashed on his face, and hot oil and wax from the lamps applied to his chest. He heard the crack of the breaking water jar, and felt the pieces of hardened clay fall into the grave. Finally, still lying immobile in the ground, he counted seven handfuls of earth taken from the crossroads, the cemetery, and the forest landing on his shroud. The mambo's sharp cry ordered him from the ground, and as the others hurriedly pushed in the loose dirt, two men tore Marcel from the grave. He was anointed again in blood, and spent the night in the sanctity of the temple.

"The blood bought back my life," Marcel concluded. "The banana never grew."

"But who did this to you?" I asked.

"I had enemies. One always does when one advances."

"The secret society?"

"No." He glanced toward Rachel. "The people."

"But you were judged."

"Yes. No. In principle, yes."

"What does that mean?"

"The plate needs the spoon and the spoon needs the plate. It is the houngan's secret."

I enjoyed seeing Marcel again, and his own story had been an unexpected revelation, but I didn't plan on doing any more business with him, at least not right away. There were several things on my mind when I returned that July, following my two months back in the States. My sponsors wanted more collections of the poison. Kline, in particular, was worried that Marcel Pierre, prodded by my brash tactics, might simply have rounded up the half-dozen most toxic ingredients at hand and improvised a preparation. I didn't share his doubts, for not only did I trust Marcel and find our evidence convincing, I also noted that the principal ingredients in his powder were the same as those in the "trap" that had brought down Clairvius Narcisse, according to his cousin. Besides, the pharmacological point had been reasonably established in lab tests.

Nevertheless, I did recognize the need for more samples. But while my backers still sought evidence of a single chemical that might explain zombification, I had become more and more impressed by a people who shared no such obsession with rational causality. I wanted to know *the magic;* I wanted to know what it meant, especially to its victims. And if the poison explained how a person might succumb, I now wanted to know why that person was chosen.

My first new lead came forty-eight hours after my visit to the Eagle Bar—through a contact in the capital provided by one of Max Beauvoir's employees, a Jacques Belfort. Beauvoir had never actually hired Jacques. Several years ago he just turned up at the gate of the Peristyle de Mariani, and it happened that Beauvoir had an errand to run. Jacques offered to do it, then he returned and did another. Gradually, errand by errand, he worked his way into the life of the hounfour. Now he appeared each morning at eight, not to set about any prescribed work, but to wait for special opportunities, knowing he could get anything done. Jacques has many wives, and one of them was from Petite Rivière de Nippes, a small fishing village in the south. There, Jacques told me, she knew a mambo who might be able to put us in touch with the ones who could make the powders, both the poison and the reputed antidote, and to whom she would be prepared to take us.

The mambo lived some distance from the coast, on a small knoll rising out of a sad landscape: fields of stones, barren trees and bushes that served for neither grazing nor firewood. It was another of Haiti's many faces, a place to find hunger.

The hounfour was deserted save for two patients awaiting treatment, and an old woman who was caring for them. Sheltered from the sun by a grass mat sat a beautiful girl, with deep almond eyes and eyelashes thick as fleece. She was dying from tuberculosis. Beside her was a small boy. The joints of mother and son stuck out like knots. Lying across from them near the entrance to the bagi was a middle-aged man whose lower leg was infected with elephantiasis.

Despite their obvious suffering, there was nothing forlorn about these people. The man laughed and welcomed us buoyantly, and the old woman, having made us comfortable in the shade, hurried to her hearth to bring coffee. The stricken man, as if embarrassed by the lack of preparation, hobbled forward and offered a share of his plate of food. I accepted and ate slowly, deliberately, hoping that he would sense my gratitude. There was so much I wanted to say, but out of respect I simply passed the plate on. It was not surprising to see such sickness in the hounfour, which is, after all, a center of healing. But to encounter such generosity and kindness in the midst of such scarcity was to realize the full measure of the Haitian peasant.

We remained with them for much of the afternoon. The mambo never showed up, but Jacques had unlimited patience, and his wife kept busy spreading the news of the capital. Some time around four, Mme. Jacques ran dry and suggested we return to Petite Rivière de Nippes and look up the mambo's son, a houngan with the name of LaBonté, "kindness." We found his hounfour without difficulty, and settled in for another long wait. Finally, well after dusk, LaBonté appeared and led us into the outer room of his temple, a small chamber with a single dusty bulb hanging from a cracked ceiling. He claimed to know nothing about zombi powders but offered instead a wide range of benevolent preparations that would increase our love, wealth, and fertility.

Rachel was adamant. "No. It is only one that we want."

LaBonté countered by suggesting that we purchase a charm, which might accomplish whatever it was we needed the powder for. Throughout this rambling conversation, Jacques stood quietly by the door in his polished shoes and pressed linen, continually wiping his brow and chest with his handkerchief. His wife sat beside him, scrutinizing LaBonté. Suddenly she stepped between him and Rachel.

"Listen," she said impatiently, "it's simple. The *blanc* wants to kill someone. If you can't give it to him, we'll go somewhere else." She turned abruptly, took Rachel by the arm, and made for the door. La-Bonté got there ahead of her.

"Something might be arranged," he said quietly. "But I will have to speak with some people. Come back tomorrow."

The jeep broke down the next day, so we were delayed a day getting back to the fishing village. The hounfour was empty, and it was an hour before we located LaBonté. His welcome marked by suspicion, he called for three associates.

"You were told to be here yesterday," he began once we were all together. "You best remember one thing. We can be as sweet as honey, or as bitter as bile. That said, we may begin the business." It was an unexpected and, for a Haitian, uncharacteristic concern with punctuality.

To prove the efficacy of their poison, one of his associates, named Obin, offered to test a prepared sample on a chicken. An amusing interlude followed as we tried to locate a chicken healthy enough for the test. Mme. Jacques categorically refused the first four, claiming that a gust of wind would blow any one of them over. Finally, she accepted a young robust rooster.

LaBonté maneuvered Rachel and me into the inner sanctum of the temple. There were no windows; the only light was a narrow beam of sun that pierced the thatch roof. One by one the others came, passing briefly through the light and disappearing into the dark. The last figure closed the door behind him and took his place in our midst. A match was struck and the flame reached forward to light first one, then two more candles. Their soft glow revealed the points of a cross, beyond which sat the houngan. LaBonté lifted his hands to the altar and began a prayer for our protection. One of the men passed around a basin containing a pungent solution and instructed us to rub the potion into our skin. Once LaBonté was satisfied that we were safe, Obin sprinkled a small amount of the poison in a corner of the chamber. LaBonté lifted a clay jar of water from his altar and told me to pour a portion of it down the rooster's throat. Moments later, Obin took the bird from my lap, placed it on top of the poison, and covered both with a rice sack.

From an obscure corner two voices, one gruff and the other strangely sensual, joined in a sonorous chant that filled the chamber. Beside me, the man who had passed the basin started to grate a human tibia. Sweat came to his brow, moistening the satin kerchief wrapped around his head as he too began to sing:

> *Make the magic Papa Ogoun, Oh!*
> *Make the magic Gran Chemin, Ogoun,*
> *That which I see, I can't talk about.*
>
> *Let me go,*
> *Let me go, people!*
> *Let me go.*
> *Rather than die unhappy,*
> *I'd rather die a young man.*
> *Let me go.*
>
> *I am Criminel,*
> *I won't eat people anymore.*
> *The country has changed*
> *Criminel says*
> *I won't eat people anymore.*

With the blessing of the songs and his asson, LaBonté selected and sanctified the bottle that would hold the vital potion that would protect me when I administered the poison. I named my intended victim, and he whispered it to the bottle. A machete cracked three times against a stone. Obin pulled four feathers from the rooster's wing and instructed me to tie them in the shape of a cross while invoking their blessing for my proposed work. The machete rang out once more. Obin led me to the cross, instructing me to make a small offering. I placed a few coins on the ground. Then, as I knelt, he inverted a bottle of clairin, causing it to bubble in a peculiar manner—a certain sign, I was told, that my desires would be fulfilled. A match dropped into the bottle exploded into flames, which for an instant illuminated the entire enclosure.

Mme. Jacques accompanied one of the men as they took the rooster to the seat to bathe its left foot. As soon as they returned, Obin threw sulphur powder into a flame, casting sparks with tails of acrid smoke that shot to all corners of the room. Then he released the rooster.

Meanwhile, the man in the satin kerchief had ground up the wood of cadavre gâté, one of the most important of vodoun's healing trees, and mixed the dust with bits of decayed matter from a human cadaver, including the shavings of the leg bone. LaBonté placed this powder in my protection bottle, adding white sugar, basil leaves, seven drops of rum, seven drops of clairin, and a small amount of cornmeal. Then he rasped a human skull and added other materials provided by one of the men, who lived by the cemetery. LaBonté handed me three candles, three powders, and a packet of gunpowder. He told me to knead the powders into the soft wax before braiding the candles into one. A third time the blade of the machete fell on the stone, harder now, and the edge of the blade scattered sparks.

The spirits answered, mounting first the man with the kerchief, then Obin and LaBonté. LaBonté filled my protection bottle and held it to my lips, piously encouraging me to drink and breathe. This I did. The spirit led me and my companions out of the bagi, into another room where he ordered us to undress. One by one, beginning with myself, we were bathed. The spirit bound my head in red cloth, and as I stood naked in a large basin of herbs and oils he cleansed my skin, with broad soothing strokes, using the rooster as a sponge. The energy of the bird would pass to me, he promised, and by the end of the bath it would lose the breath of life. Rachel followed me into the basin, and then Jacques, and by the time Mme. Jacques was clean, the rooster lay on the ground, flaccid and quite dead.

"It is good," Mme. Jacques said. "In Port-au-Prince the basin is terrible. Here you smell of beauty, even though you are about to kill."

Now that we were safe, the spirit directed us back into the bagi for the preparation of the actual poison. There was a new song invoking Simbi, the patron of the powders.

> *Simbi en Deux Eau*
> *Why don't people like me?*
> *Because my magical force is dangerous.*
>
> *Simbi en Deux Eau*
> *Why can't they stand me?*
> *Because my magical force is dangerous.*
> *They like my magical force in order to fly the*
>   *Secret Society.*
> *They like my magical force in order to be able*
> *To walk in the middle of the night.*

There were four ingredients: one was a mixture of four samples of colored talc, another was the ground skins of a frog, the third was gunpowder, and the fourth was a mixture of talc and the dust ground from the dried gall bladders of a mule and a man. There was no fish, and no toad.

I glanced quickly toward the others, first Rachel, then Jacques. Both sat still and unchanged, but Mme. Jacques had shed her years like water. Her dress fell away from one shoulder, and she had crossed her legs so that a bare foot rested high on her thigh. From a wiry, grim peasant woman, she had become sultry and seductive. Her lips squeezed a cigarette, but it was the wrong way around—the lit end sizzled on top of her tongue.

Her husband caught my stare. "There is no problem," he confided. "Often when she is taken by the spirit she rubs the juice of the chile pepper on her vagina. Listen!"

The body that had been Mme. Jacques was singing. "We are assembling, we are near the basin, we are going to work. We don't know how it will be but we shall do the work." The linked phrases of this high, plaintive wail merged with the rattle and whistle and bells in an ominous cacophony unlike any sound I had heard in a vodoun hounfour.

Then, speaking with a voice that was not hers, she demanded a second poison. Without argument one of the men brought forth a small leather pouch and emptied the contents into a mortar.

"These are the skins of the white frog," the spirit intoned. "The belly of your victim will swell, and let them cut into it, it will bleed a river of water."

I lifted one of the skins from the mortar and held it close to the candle. Even I could recognize it as that of the common hyla tree frog. Small glands beneath its skin secrete a compound that while irritating is hardly toxic.

To administer the poison, I was told, it was critical for both my own safety and the success of the work that I follow instructions precisely. On the night of the deed, I was to light the braided candle and hold it up before the evening star and wait until the sky darkened. To cast the death spirit, I would first have to beseech the star saying:

> By the power of Saint Star,
> Walk, Find
> Sleep without eating.

Then, having saluted a complex sequence of stars, I was to place the burning candle in one of two holes dug beneath my victim's door. Next, I was to drink from my protection bottle to imbibe the power of the cemetery. To set the trap I had merely to sprinkle the powder over the buried candle, staying carefully upwind while I whispered the name of my intended victim. Once the fated individual crossed over the poi-

son, death would be imminent. As a final precaution, the spirit warned
me to sleep with the cross of feathers beneath my pillow. That way the
power of this ceremony could continue to shield me. With these final
words, the spirit left.

"With this your enemy will fall," Obin assured me as we were
about to leave the hounfour. In his hand he held a small jar that contained the second preparation.

"And to make him rise again?" I asked, still clinging to the notion
of an antidote.

"That is another magic. For what you have there is no treatment.
It kills too completely."

"And the other powder?" Mme. Jacques asked Obin.

"It's the same," he said. "With these you will kill. Is that not what
you want?"

"There's more."

"What you have been given is *explosive*. Both powders shall leave
your enemy but one ark, the earth that shall take him."

"I want his body," I said.

"For that you must return."

"When?"

"When he is dead and you are ready."

It was dusk and a young moon hung over the sea, but it was still
hot. Jacques cracked open a bottle of rum, and we drank as we walked
away. For a while no one spoke. Our clothes clung to our skin, and we
smelled of the market—a combination of sweat, jasmine, and rotting
fruit. The fishermen were out, two rows along the shore, and we
watched the coils grow at their feet as they hauled in the ends of the
great semicircles of net, which closed on piles of flotsam.

"Of course," Mme. Jacques explained once we had reached the
jeep, "there are dozens of powders. They walk in different ways. Some
kill slowly, some give pain, others are silent."

"And the ones we bought?"

"They *carbonize*. But it is the magic that makes you the master."

"What of the others?"

"It is easier. In food. Or they prick the skin with a thorn. Sometimes they place glass in the mortar. It is a matter of power. If you
want to learn the powders, you best walk at night." Mme. Jacques
accepted the bottle of rum. "But now," she said, "you have known the
face of the *convoi*. The society has touched you."

"How does she know these things?" Rachel teased, wrapping her arms around Jacques's neck.

"Oh!" he cried, gagging on a swallow of rum and collapsing into uproarious laughter. "How might she know! They call her Shanpwel. Those men are her cousins. Obin is the president. She is the queen!"

Two men were waiting for me at Beauvoir's that evening. One was the chief of police of a city in the north. The other I could have recognized by sound alone—by the peal of throaty laughter filtered through a thousand cigarettes until it had the edge of a rasp. He was the same man who had been waiting for us, with three others, on the night of my second day in Haiti, when we returned from Marcel Pierre's with the bogus preparation. Then he had poured the sample onto his hand derisively. This time I learned his name—Herard Simon. He was equally blunt now.

There were at least four preparations that could be used to make zombis, and for the proper amount of money I could obtain them all. It was a substantial sum. I called New York from Max Beauvoir's phone and received instructions to buy one powder, and if it worked on laboratory monkeys, I could return to purchase the others. I returned to Simon and halved the price. He agreed, and I gave him a deposit. He told me to meet him in the north in three days, then he left.

Our meeting had lasted scarcely longer than the time of my long-distance call, but my impression of Herard Simon carried well into the night, until it kept me from sleep. Outwardly, he seemed calm, almost sluggish, for long ago, the angles of his body had disappeared beneath a mountain of flesh. But like the Buddha he resembled, his corpulence had a purpose; beneath it there was something at once terribly wise and terribly savage, like the soul of a man who has been forced to kill. Nobody told me until much later, but already that night I knew: in meeting Herard Simon, I had met the source.

"It was hard-hit," Rachel said. Her words took in all the confusion of dust and rubble that was Gonaives. The power was out, and in the darkness the city looked unnatural, its buildings half-abandoned, yet its streets alive with people. In the market the drifters and sellers huddled around small fires, their children in clumps. Everyone seemed to be living outside, like survivors camped atop a ruin.

"They closed the port," Rachel said.

"When the road went through?"

"Before. Duvalier wanted everything in the capital."

"So all the business left?"

She nodded. "Turn here," she said suddenly, and we veered onto a gravel road riddled with potholes. "It used to be in Gonaives that if you were black, the mulattos wouldn't sit beside you."

"Duvalier changed that?"

"The revolution did. Now there's hardly any mulattos left here."

"That's convenient," I mumbled.

"What?"

"Nothing. Say, have we been here before?"

"You don't remember?"

Then I did. Even in the darkness you couldn't miss the mermaid swimming along that blue-and-green wall. The woman who ran the Clermezine nightclub, and who had expressed such a low opinion of Ti Femme, was the wife of Herard Simon.

I swung abruptly into the short drive, and for an instant the beam of the headlights froze the same amorphous group of idlers against the gate of the compound.

"Hélène's away on pilgrimage, but Herard's probably here." Rachel started to say something to the men sitting around, but then stopped, hesitated, and, grabbing her cigarettes, stepped out of the jeep. She drew several of the youth toward her, and then quite deliberately asked one of them for a light.

Once back in the jeep, she said, "He's not in. They say he had business."

"Where would that be?"

"Anywhere. Maybe in town. Sometimes he hangs out by the waterfront. What shall we do?"

"Wait."

I pushed open the door, propped a foot on a hinge, and settled back. Some of the youths gathered around. You could tell they were thirsty, so we shared what was left of the rum. It was good to watch the bottle pass around. That's a special thing about Haiti—everyone loves to drink, but you never see anyone drunk.

We chatted away for a while, but gradually they drifted back to the gate, finished with us and ready to sink back into their nightly routine.

"Strange," Rachel said once we were alone. "Did you notice the one on the left, the one that lit my cigarette?"

"You recognized him?"

"Not at first. Then I remembered. He was in L'Estère with Narcisse's sister. I'm certain."

"What's he doing here?"

"He's from here. The question is, what was he doing there?"

It was about nine when a slim figure slipped up to the side of the jeep and startled Rachel. Without identifying himself, he said quietly, "The commandant is waiting at his place." Then he walked away.

"The commandant?" I asked as we pulled out of the drive.

"Everyone calls him that. He used to be head of the militia, the VSN." She used the official acronym for the Ton Ton Macoute.

"Of Gonaives?"

"No, all the Artibonite."

"When was that?"

"Right at the beginning. He's retired now, but he still runs things. My father says that his people have been watching us since you arrived."

Herard Simon didn't have a lot to say. He sat alone, on the porch of a simple dwelling, absentmindedly brushing the flies away from his face. We shook hands at my initiative. Sometimes, when strangers meet because they must and nothing is said, the silence is honest. But here it wasn't. It was a statement of his authority, and I had to struggle against an urge to crack it. When he finally spoke, it was with a voice that placed a shell of double meaning on every word.

"What do you care of zombis?"

"I'm curious."

"Curious? You pay all this money because you're curious?"

"Someone else pays."

"The *juifs* [Jews]. Of course, you are not one of them. They send you because they won't do the work. And who will make the money?"

"From what we have arranged, it seems that you will," I said, ignoring his swipe at Kline and my other backers. It went on like this for some time, he asking all the questions, baiting me with his knowledge of my past.

"The *blancs* are blind," he said, "except for zombis. You see them everywhere."

"Zombis are a door to other knowledge," I said.

"To death and death alone!" he exclaimed in a suddenly strident voice. "Vodoun is vodoun, zombis are zombis." His calm returned just as quickly. "So," he said, "you have seen Narcisse."

"Yes, and his family too."

"Well?"

"He lives."

"Yes, one who comes from the ground can be quite normal. But tell me, *blanc*, if you were a woman, would you ask him to dance?" That cracked him up, and once again I heard that rasping laugh.

"They say he's got a lawyer, and he's trying to get back his land to work it again." This made him laugh again, even harder. "This man Narcisse is half-intelligent. As if he can get protection in the capital from his own people." He turned to Rachel. "Beauvoir! This is enough. Bring yourself and this *blanc malfacteur* back in the morning. Then we will begin the work."

That night, while Rachel and the others at the nightclub slept, I lay in bed struggling for an answer that would explain it all. It hadn't surprised or worried me that Herard Simon knew so much about our activities; they hadn't been secret, and there were any number of obvious sources of information. What concerned me was the man himself. I couldn't let him be. And I had barely met him, that was the extraordinary thing. He had that kind of presence, a charisma hot to the touch. There was something frightening about him, a latent violence that was both ancient and tribal. It seemed as if he bore within him the exploding energy of an entire race; as if his skin, stretched so thin over his massive body, lay ready to split, to release some great catastrophe of the human spirit. He exuded power. I felt it that night, as I had when we first met, and I would experience it again the following morning.

The party from Petite Rivière de l'Artibonite arrived just before noon, but the reputation of their region preceded them. Among other things, I had been told that there was scarcely a bone left in their public cemetery. They were five. The two riding in the cab of the pickup, one wearing an army uniform, had the bearing of officials. The three in the back had the look of mountain peasants. The scalp of one of these was dotted with furry patches—an occupational hazard, I was told, of the malfacteur, the one who grinds the powders.

Herard Simon directed us into the outer chamber of his hounfour. The negotiations were layered with go-betweens, but with no introductions. Herard spoke first, with a few measured words that seemed to secure each person to his will, establishing himself as the principal broker, and then said little. The military official hovered somewhat paternalistically over the peasants, but they spoke for themselves, in a heavily accented Creole that betrayed their roots deep in the mountains. They had a zombi, they claimed, and they also offered to prepare a sample of the poison. The discussion flared with proposal and counterproposal, while their hairless leader tossed off figures in great flurries of bravado, as if the mere mention of such astronomical sums might, like a charm, make them come true. His two partners clearly enjoyed the whole business; they rose up and down on their haunches, urging him on. When I sliced the price, they reacted with pious indignation.

Throughout all this Herard Simon sat silently near the wall, leaning forward on a wooden chair, resting his weight on his knees, virtually immobile. A menthol cigarette burned in his motionless right hand. His very indifference kept command of the conversation. Finally, perhaps tiring of the inconclusive banter, he lifted both his arms and turned to the soldier.

"Where is it?" he asked.

"They have it with them."

"Then bring it forward. Perhaps a zombi will loosen the *blanc*'s purse."

The one in uniform said something to the leader of the peasants. He started to argue, but stopped and, reaching into a dusty bag, pulled out a small ceramic jar wrapped in a red satin cloth. Herard started to laugh, but his laughter had the edge of anger.

"Fools!" he cried. "Not *that*. They want the flesh."

The significance of his words was lost for the moment in the blur of his rage. Scattering invectives, he banished all of us from the hounfour. The peasants fled, the officials foundered, and with total contempt Herard made his way back across the courtyard to his house.

He was laughing more jovially by the time we caught up with him. "Imagine, *blanc*," he said to me, "they brought you a zombi astral because they didn't think you'd be able to get a zombi of the flesh through immigration!"

"Wait a minute . . ." I started to say.

The expression on Herard's face told me he was surrounded by idiots and I was one of them. "Three days," he answered, "return in three days." Again I started to say something, but his thick fingers passed once before his face to cut me off.

Two types of zombis, I thought. That changed everything.

In the summer in Haiti the spirits walk, and the people go with them. For weeks in July the roads come alive with pilgrims, and we followed them.

Leaving Gonaives, Rachel and I drove north across the mountains to the lush coastal plain, calling first at the sacred spring and mudbaths of Saint Jacques, and then moving on to the village of Limonade and the festival of Saint Anne. Here they had gathered, literally thousands of them dressed in the bright clothes and colors of the spirits, fused in hallucinatory waves that flowed across the plaza.

The seething edge of the throng enveloped us even as we stepped from the jeep. We were carried, flesh to flesh, by the collective whim of the crowd. It was like being pushed through the stuffed belly of a beast, and soon we were ploughing through the throng to the nearest refuge, the stone steps of the church standing firm like a jetty above the madness.

Our senses numbed, we entered the church and were well inside the nave before we realized what was going on. It was the Mass of the Invalids, and at our feet lay the most diseased and wretched human display imaginable. Lepers without faces, victims of elephantiasis with limbs the size of tree trunks; dozens and dozens of dying people, collected from the length and breadth of the country to seek alms and redemption at the altar of this church. It was a scene of such singular horror, we could think only of escape.

Rachel stepped ahead of me toward an open door, and then gasped. There in the shadow of a cross, her head covered by a black shawl, was a single woman, and draped across her legs was her daughter, a teenage girl whose shattered legs crossed like sticks. Her skin was jet black and her head a grotesque melon, so swollen with disease that you could see the individual follicles of hair. It was a sight so terrible that we could not pass. We turned back to wade through the brown-frocked beggars carpeting the front of the church, and as we passed they tugged at our clothes. There was nothing for them, and the real horror of the moment was less their condition than our fear.

Then, on the steps of the church, the scene turned into an epiphany. A healthy peasant woman, dressed in the bright-blue-and-red solid block colors of Ogoun, the spirit of fire and war, swirled through the beggars possessed by her spirit. Over her shoulder was slung a brilliant red bag filled with dry kernels of golden corn. She twirled and pranced in divine grace, and with one arm stretching out like the neck of a swan she placed a small pile of corn into each of the begging bowls. When she was finished, her bag empty, she spun around to the delight of all and with a great cry flung herself from the steps of the church. Rachel and I watched her flow into the crowd. Wherever she went the people backed away, that Ogoun might have space to spin. Our eyes followed her until she was gone, and then without speaking we dropped back into the crowd.

Our travels during the rest of that awesome day took us back across the plain to the old colonial capital of Cap Haitien, a gentle place whose warm texture belies a bloody history. At a house built on a ruin with material taken from the sea, we rang the bell of Richard Salisbury, known throughout Haiti as the British consul.

Salisbury, from what I had been told, had an enthusiastic interest in vodoun, and we hoped that he might provide some information concerning the time Narcisse had spent in Cap Haitien immediately following his release. At first there was no response, but after a second ring, the shutters of a second-floor window flung open, and the hot afternoon sun fell harshly on the etiolated body of a middle-aged man. He had just woken up.

Salisbury received us on his veranda in the shadow of an enormous Union Jack. With his handlebar moustache, peppermint complexion, and extended belly wrapped carelessly in a silk smoking jacket, he was a character straight out of Somerset Maugham. As it turned out, his knowledge of vodoun was superficial, and in fact he had nothing to do with the British government. An accountant whose meager investments in a local sugar mill had, until recently, allowed him to live royally in Haiti, he was less a diplomat than a metaphor for the demise of the empire. Salisbury now faced a major personal crisis, and we, unfortunately, were in a position to hear all about it. Corrupt partners and a depressed international market had bankrupted the mill, and he had no choice but to return to England. There he would face the life of any other middle-class accountant, riding the subway to a repetitive, meaningless job. Returning was the last thing he wanted to do, and now given the opportunity he turned to Rachel and asked quite desperately for advice. The sight of this grown man, this European whose attributes were a bit of capital and the false status once afforded to the color of his skin, beseeching a young Haitian girl, walked the fine line between comedy and tragedy.

Close to dusk, we managed to extract ourselves from the problems of Richard Salisbury and made for a coastal beach just east of the city. There beneath the palms, with the sun turning copper, we finally rested. The day had started off in the house of Herard Simon in a confrontation with the poison makers, had led to the mudbaths of Saint Jacques and the horror of Saint Anne, and then to the anachronistic Richard Salisbury. Now it ended on a pristine, tranquil Caribbean beach. I looked past the trees and heard the shrill, incoherent cry of the gulls as they swept and snapped in the waves. Down the shore there was a pair of cormorants, pelicans too. The luxury of wild things, lush and unreasonable. And in the water, Rachel, swimming like a dolphin.

Three days later, as previously arranged, we met again with Herard Simon in Gonaives. He was where he could be found every night, near the waterfront by a dilapidated movie house, his finger on the pulse of the street. His greeting this time was surprisingly cordial. Apparently my status had shifted somewhat—in what direction, I was not certain— for in place of the anonymity of *"blanc"* he now addressed me as his "petit malfacteur," his little evildoer. Herard began by emphasizing that as a houngan he had no interest in zombis; they were nothing, he insisted, compared to the profound lessons of the vodoun religion. For business reasons, however, he had made the necessary arrangements. On the morrow, he promised, one of his contacts would begin to prepare the zombi powder.

"And malfacteur," he said just before we left him, "with what I give you your monkey will go down, it will not come up, and it will never again wag its tail."

It took a full week to make the poison.

First Herard, as houngan, prepared the antidote, which, not surprisingly, contained a plethora of ingredients, none of which had significant pharmacological activity. It consisted of a handful of bayahond leaves (*Prosopis juliflora*), three branches of ave (*Petiveria alliacea*), clairin, ammonia, and three ritualistically prepared lemons. As in the case of the reputed antidote prepared by Marcel Pierre, there was no evidence that it could chemically counteract the effects of any poison.

The actual poison did have potent constituents, and critically the ingredients overlapped in significant ways with those used at Saint Marc. Herard's man distinguished three stages or degrees to the preparation. During the first a snake and the bouga toad (*Bufo marinus*) were buried together in a jar until "they died from rage." Then ground millipeds and tarantulas were mixed with four plant products—the seeds

of tcha-tcha (*Albizzia lebbeck*), the same leguminous species added
by Marcel Pierre; the seeds of *consigne* (*Trichilia hirta*), a tree in the
mahogany family with no well-known active constituents; the leaves
of *pomme cajou*, the common cashew (*Anacardium occidentale*); and
bresillet (*Comocladia glabra*). The last two plants are members of the
poison ivy family, and both, especially bresillet, can cause severe and
dangerous dermatitis.

These ingredients, once ground to powder, were placed in the jar
and left below ground for two days. Then, at the second degree, two
botanically unidentified plants known locally as *tremblador* and *des-
membre* were added. For the third and final degree, four other plants
capable of causing severe topical irritations were mixed in. Two were
members of the stinging nettle family, *maman guêpes* (*Urera bacci-
fera*) and *mashasha* (*Dalechampia scandens*). The hollow hairs on the
surface of these plants act as small hypodermic syringes and inject a
chemical similar to formic acid, the compound responsible for the pain
of ant bites. A third plant was *Dieffenbachia seguine*, the common
"dumbcane" of Jamaica. In its tissues are calcium oxalate needles that
act like small pieces of glass. The English name derives from the nine-
teenth-century practice of forcing recalcitrant slaves to eat the leaves;
the needles, by irritating the larynx, cause local swelling, making
breathing difficult and speaking impossible. The fourth plant, *bwa piné*
(*Zanthoxylum matinicense*), was added because of its sharp spines.

The addition of these irritant plants recalled Marcel Pierre's use of
Mucuna pruriens, the itching pea. It was of interest that several of these
additives could produce such severe irritation that the victim, in
scratching himself, might quite readily induce self-inflicted wounds. I
knew from the results of the laboratory experiments in New York that
the powder, though topically active, was particularly effective if applied where the skin had been broken. Mme. Jacques had suggested that
ground glass might be used. And of course I had reason to believe that
when the powder was administered the skin of the victim was quite
deliberately broken. It had been stated that the powder might be applied more than once, so it was possible that the irritant plants directly
increased susceptibility to subsequent doses.

It was the list of animals added at the third degree that gave me the
greatest sense of satisfaction. Tarantulas of two species were ground
with the skins of the white tree frog (*Osteopilus dominicencis*). Other
ingredients included another bouga toad and not one but four species

of puffer fish (*Sphoeroides testudineus, S. spengleri, Diodon hystrix, D. holacanthus*). Thus, in common with the poison prepared by Marcel Pierre, we had the toad, the puffers, including the sea toad, and the seeds of *Albizzia lebbeck*.

Over the course of that week our relationship with Herard Simon warmed considerably. There was no one dramatic turning point, as there had been in the case of Marcel Pierre; Herard was far too clever and wary for that. Rather, it was a number of small incidental things that he appreciated—the fact that we drank from his well, that we shared his plate, that we curled up beside him on the stony ground.

For one reason or another, in time, he chose to loose three critical pieces of information. First, he gave me the names of four preparations used to create zombis—*Tombé Levé, Retiré Bon Ange, Tué,* and *Levé*—and though he refused to describe the specific formulae, he did offer the facts that one killed immediately, another made the skin rot, and a third caused the victim to waste away slowly. He also commented that these virulent preparations had one ingredient in common—the crapaud de mer, the most toxic of the puffers found in Haitian waters.

Secondly, Herard told me that the best powders were made during the hot months of the summer, and were then stored and distributed throughout the year. At the same time, he cautioned that some of these were excessively "explosive," that they killed *too* completely. From the research I had done in Cambridge I knew that levels of tetrodotoxin within the puffer fish are not consistent. They vary not only according to sex, geographical locality, and the time of the year, but from individual to individual within a single population. A puffer from Brazilian waters, *Tetrodon psittacus*, for example, is only poisonous in June and July. Among Japanese species toxicity begins to increase in December and reaches a peak in May or June. The species used in the zombi preparations show a similar pattern—*Sphoeroides testudineus*, the sea toad, is most toxic in June, precisely the time when Herard said the poison was strongest.

Finally, Herard told us that when the zombi is taken from the grave it is force-fed a paste, with a second dose administered the next day when the victim reaches its place of confinement. The ingredients of the paste were three: sweet potato, cane syrup, and, of all things, *Datura stramonium*.

It was a startling piece of information. Since the beginning of the investigation, the role of this potent psychoactive plant, so suggestively named the zombi's cucumber in Creole, had eluded me. Now a dozen incongruous facts crystallized into an idea. So far the search for a medical antidote for the zombi poison had turned up nothing of pharmacological interest. Each zombi powder had its locally recognized "antidote," but in each case the ingredients were either inert or were used in insufficient concentration. Moreover, there was no consistency in either their constituents or the means of preparation between the various localities. Now, with Herard's revelation, I had reason to believe that if there was an actual antidote, it was the zombi's cucumber!

Tetrodotoxin is a most peculiar molecule. No one is exactly sure where it originated. Generally, such specialized compounds pop up just once over the course of evolution, and as a result are only found in closely related organisms, derived presumably from a common ancestor. For the longest time, tetrodotoxin appeared to be isolated to a single family of fish. Then, to the surprise of biologists, it turned up in an amphibian, the California newt, a totally unrelated creature. Subsequent research found it in the goby fish from Taiwan, atelopid frogs from Costa Rica, and the blue-ringed octopus from the Great Barrier Reef of Australia. Such an erratic distribution suggested to some scientists that the toxin originated lower in the food chain, perhaps in a small marine organism.

Within the puffer fish themselves, toxicity is correlated with the reproductive cycle and is higher in females, but the remarkable variability in toxin levels among separate populations of the same species has prompted similar suggestions that the concentration of the toxin may be linked to food habits. Puffer fish grown in culture, for example, do not develop tetrodotoxins, and it is possible that the puffer fish, in addition to synthesizing tetrodotoxins, may serve as transvectors of either tetrodotoxin or ciguatoxin, a different chemical that originates in a dinoflagellate and causes paralytic shellfish poisonings. The symptoms of ciguatera poisoning are similar to those of tetrodotoxin and include tingling sensations, malaise, nausea, and digestive distress, with death resulting from respiratory paralysis.

In Australia, still today as throughout their history, the aborigines have a very strange plant, actually a tree, that they call *ngmoo*. They carve holes in its trunk and fill them with water, and within a day have an interesting beverage that produces a mild stupor. The branches and leaves, they have also learned, when placed in standing water quite effectively intoxicate eels, forcing them to surface where they can be killed. Knowledge of the remarkable properties of this plant found its way north to New Caledonia. There the native inhabitants discovered that the leaves could be used to make an effective antidote to ciguatera poisoning, an observation that modern science has confirmed. The plant is *Duboisia myoporoides,* and like many members of the potato family it has a number of potent chemicals including nicotine, atropine, and scopolamine. There is no known medical antidote for tetrodotoxin, but in the laboratory it has been shown that, as in the case of ciguatera poisonings, atropine relieves certain symptoms.

Datura stramonium, like its relative from New Caledonia, contains atropine and scopolamine, and hence could be serving as an effective but unrecognized counteragent to the zombi poison.

The investigation had come full circle. Ironically, the plant I had originally suspected to be the source of the drug that allowed an individual to be buried alive turned out to be, if anything, a possible antidote, which, at the same time, was instrumental in actually creating and maintaining the zombi state. For if tetrodotoxin provided the physiological template upon which cultural beliefs and fears could go to work, datura promised to amplify those mental processes a thousand times. Alone, its intoxication has been characterized as an induced state of psychotic delirium, marked by disorientation, pronounced confusion, and complete amnesia. Administered to an individual who has already suffered the effects of the tetrodotoxin, who has already passed through the ground, the devastating psychological results are difficult to imagine. For it is in the course of that intoxication that the zombi is baptized with a new name, and led away to be socialized into a new existence.

Further evidence of the makeup of the zombi poison came two days later south of the capital near the town of Leogane. Several weeks previously we had established contact with a houngan named Domingue. His son, named Napoléon, was a well-known malfacteur, and he had a message for me at Beauvoir's. My meeting with Napoléon was brief, but it yielded two poisons of note. The most toxic, by Napoléon's account, was made from human remains alone. It consisted of a ground leg bone, forearm and skull, mixed with dried and pulverized bits of dried cadaver, and it was the first and only poison I encountered administered in the reputedly most traditional way. Having rubbed one's hands with the protective lotion—again a mixture of lemons, ammonia, and clairin—the killer sprinkled the powder in the form of a cross on the ground, while naming the intended victim. The recipient need only walk over that cross to be seized with violent convulsions. If the powder was placed in the victim's food, the action would be immediate and permanent.

The second preparation was a more familiar mixture of insects, reptiles, centipedes, and tarantulas. In place of the bouga toad, two locally recognized varieties of the tree frog were added. Napoléon also included the sea toad, the crapaud de mer. The onset of the poison was characterized by the feeling of ants crawling beneath the skin, precisely the way that Narcisse had described his first sensations. Besides evidence of yet another preparation based on the puffer fish, Napoléon gave further indication of the importance of correct dosage. He mentioned that the animal powder was most effective if ingested by the victim, and he cautioned that his two preparations should never be mixed. Together they would act too explosively; the victim would be too dead and would never rise again.

I left Leogane confirmed in my conclusion that tetrodotoxin was the pharmacological basis of the zombi poison, a conviction that was reinforced by subsequent collections at various localities along the coast of Haiti. Finally I felt I could let the issue of the poison rest. It was time to move on to other matters that had claimed my interest since returning to Haiti.

Herard Simon called early the next day and insisted on seeing Rachel and me immediately. We left that afternoon, picked him up in Gonaives, and drove directly to Petite Rivière de l'Artibonite. Arriving at the army caserne by dusk, we followed Herard past the sentry and into the private quarters of the commander. After sending the commander on an errand, Herard asked his orderly to bring us food and a bottle of rum. Then we waited. Past a broken shutter, the rain poured down, and inside a long row of cells I could just make out the prisoners clinging to the bars, trying to keep their feet out of the water.

Herard is not a man who likes questions. I tried three. I asked first about the astral zombi that had been brought by the mountain peasants.

"In that bottle was the soul of a human being," Herard replied, "the control of which is an ominous power. It is a ghost, or like a dream; it wanders at the command of the one who possesses it. It was a zombi astral captured from the victim by the magic of the bokor."

"What about the poison you gave me?"

"A poison kills. You put it in food. I gave you no poison."

"But the powder?"

"Yes. Powder is powder, poison is poison. The powder is the support of the magic. Only the truly great work magic alone. Small people pretend, but watch and you'll see the hand of powder. There are some, and I know them, who can stand in front of an army and throw a spell on anyone."

"To steal the soul?"

"What else? If you want the flesh to work, you can't fool with a little powder dust on the ground. You take a bamboo tube and blow it all over him and you rub it into the skin. Only then shall the zombi cadavre rise."

With the realization that there were two kinds of zombis and two means of creating them, many of my loose ends came together. Clearly what LaBonté and Obin had offered was a powder that, embued with sorcery and triggered by a magical act on my part, would kill my enemy. That was what Mme. Jacques had been telling me—"the magic makes you the master." It was, in effect, voodoo death, the Haitian equivalent of the aboriginal practice of pointing the bone. It would not have been particularly effective within the mindset of my society, but that wasn't their problem, it was mine.

There was another possibility. If your spiritual force was strong, Marcel Pierre had told us, you could resist a spell, but the powder promised to get you anyway. Mme. Jacques mentioned powders that were rubbed into the skin or placed in wounds; she had talked of glass ground in the mortar, of the skin ripped by a thorn. These had to be the pharmacologically active powders that allowed the zombi to rise. It was what Narcisse had said. On the Sunday before his death, they took him before the basin and pricked his skin, and the water turned to blood! Herard was right. If you wanted a zombi of the flesh to rise, a little powder sprinkled on the ground would never do!

My third question was interrupted by the arrival of the president of the local secret society. At that point the army commander, who had returned while I had been questioning Herard, was reduced to serving us rum and food. Herard discussed the possibility of obtaining a zombi cadavre for medical study. It was not until the middle of the night that we finally agreed on a price. Herard arranged to meet again with the president and his people the next afternoon. As we drove out of town, I attempted to arrange a time to meet the following day.

"No. No," Herard laughed, "we don't return tomorrow. I just wanted to measure their force."

"But how will they bear the insult?" I asked, recalling the somewhat ominous warning of LaBonté at Petite Rivière de Nippes. "Are not the societies everywhere?"

"Yes, they are powerful," he agreed. "That's why you had to come here with someone who is stronger. I, too, represent a secret society. Mine is a society of one!"

Then, as if to emphasize that there were limits to even his own authority, he warned me that around Petite Rivière de l'Artibonite anything could happen. The road forked and he pointed to the shadows.

"Be especially careful at the crossroads. Never leave your car after dark. They move by night."

As we approached Gonaives, I felt compelled to ask a final question, the one that I myself would be asked so often. Why was he giving me all this information? He laughed, but did not respond until we reached his house. Then, just as he left the jeep, he looked back at me.

"*Mon petit* malfacteur, you are not a fool, but you still do not understand. You may gather your powders; in fact, I will give you all the powders you want. You will meet zombis, you may see zombis come from the earth, you will even think that you understand zombis. But you will never make a zombi, nor will you leave here with the magic.

"Someday when you stop asking all these questions, you will begin to see. Only then will you begin to know vodoun, and only then will you step into the path of the loa."

For all his gruff manner, Herard had provided more information concerning the poison than anyone else; yet of all my contacts in Haiti, he had expressed the most disdain for such powders, and the least interest in zombis. He was a deeply religious man, a theologian really. Sometimes when Herard said things I had no idea what he meant; at other moments his words seemed like a beacon. It was time I tried to understand the spirit world of his people.

10

The Serpent
and the Rainbow

ON JULY 16, 1843, and then again on the same day in 1881, the Virgin Mary appeared on the top of a palm tree near the village of Ville Bonheur in the rugged mountains of central Haiti. She said the world was going to end. This was most convenient for the Roman Catholic church, for the palm tree grew not far from the base of a cliff where the La Tombe River dissolves into mist, a waterfall named Saut d'Eau that had been a sacred vodoun pilgrimage site for a good many years. The Catholic priests, then as always anxious to purge the nation of what they considered a pagan cult, took immediate advantage of their good fortune. A chapel and shrine were erected, the legitimacy of the apparition was verified by church authorities, and thenceforth the miraculous event was commemorated annually with a full day of religious celebrations. But a strange twist was soon added to the saga. With increasing frequency, the officiating priests found small plates of food placed alongside the votive candles at the Virgin's shrine. Once they realized what was going on, their initial enthusiasm faded rapidly. Rather than their co-opting a traditional vodoun pilgrimage, quite the opposite had taken place. For the peasants, the apparition of the Virgin Mary was none other than Erzulie Freda, the goddess of love, and her presence was less a miracle than an expected blessing that only added to the reputation of the sacred waterfall.

The Virgin next appeared during the American occupation, and by then the clergy was prepared. A local priest dismissed the apparition as idle superstition and called upon a Marine captain stationed in the region to help suppress the worship. The Marine complied, ordering a Haitian sergeant to shoot into the blinding light. As he did so, the vision moved to another palm, and then another, until the exasperated priest ordered the trees cut down. The vision rose above the canopy, and as the last of the palms fell, it changed into a pigeon. For a moment the priest was satisfied, but just as he turned to leave word came that his house had burned down, destroying all his possessions. This calamity was only the beginning for him; within a week the priest lay dead, the victim of a paralytic stroke. The American was similarly punished, and the Haitian sergeant went insane and was found sometime later wandering alone in a forest near the village. From what the townspeople say the pigeon remained close to Ville Bonheur for several days and then flew to Saut d'Eau, where it disappeared into the iridescent mist.

The waterfall at Saut d'Eau carves a deep hidden basin from a limestone escarpment, and by the time I arrived shortly after midnight the entire vault was bathed in the soft glow of a thousand candles. Already in the depths where no moonlight could fall, huddled together, or darting in and out of the water, singing the vodoun songs or serving the many altars, were dozens and dozens of pilgrims. On all sides, people saturated with a lifetime of heat shivered and trembled, drawing in their hands against their naked skin. High above on the trail along which I had passed, other seekers had abandoned themselves to their goal and drifted along the horizon with the motion of night clouds. Overhead, beyond the crown of the towering mapou tree that hovered over the basin, the branches of heaven spread and the stars scattered as thick as blossoms in the northern spring.

Vodoun is not an animistic religion, Max Beauvoir had told me. The believer does not endow natural objects with souls; they serve the loa, which by definition are the multiple expressions of God. There is

Agwe, the spiritual sovereign of the sea, and there is Ogoun, the spirit
of fire and the metallurgical elements. But there is also Erzulie, the goddess of love; Guede, the spirit of the dead; Legba, the spirit of communication between all spheres. The vodounist, in fact, honors hundreds of
loa because he so sincerely recognizes all life, all material objects, and
even abstract processes as the sacred expressions of God. Though God
is the supreme force at the apex of the pantheon, he is distant, and it is
with the loa that the Haitian interacts on a daily basis.

The spirits live beneath the great water sharing their time between
Haiti and the mythic homeland of Guinée. But they often choose to
reside in places of great natural beauty. They rise from the bottom of
the sea, inhabit the rich plains, and clamber down the rocky trails from
the summits of the mountains. They dwell in the center of stones, the
dampness of caves, the depth of sunken wells. The believer is drawn
to these places as we are drawn to cathedrals. We do not worship the
buildings; we go there to be in the presence of God. That is the spirit
of the pilgrimage.

Having bathed in the falls, I made my way to the mapou, and
there among the buttresses and serpentine roots found shelter from the
cold, damp wind. The roar of the water dominated all other sounds,
but before long it fell away, leaving a welcome cushion of silence, the
hollow tone one imagines deaf people hear all the time.

There were two snakes, I was warned, one green and one black,
that lived at the base of the tree. If so, they left me alone. From the
edge of a fitful sleep, I sensed only the thick hide of the mapou on
either side of my face, and beneath my hands the texture of root bark.
I knew every structure within that tree, each vessel, each pore and
trichome, the placement of each stamen, and the pathways of every
drop of green blood. In botanical studies I had watched it dissected
into a thousand or more parts until each one lay isolated, a separate
hypothetical event, simple enough to be explained according to the
rules of my training. This was the legacy of my science. Each of us
chipping away at the world, doing our bit. But what was I to make of
Loco, the spirit of vegetation, the one that gives the healing power to
leaves? This was his home, and it seemed to me strangely alive and different suddenly, not a series of components but a single living entity,
animated by faith.

I caught a fingernail in the bark, and it sent chills up my back. I sat up abruptly. At my feet and all around the tree the pilgrims curled up like sheepdogs, their bones stiff and soaked in darkness. Around me in the crotch of other roots, I saw the faint glimmers of other penitents placing candles at altars that weren't there when I lay down. Hands reached forward, pressing soft wax into a fissure in the smooth bark. The flames flickered and spat smoke and kept going out. Below us all, the sheer power of the falling water.

I woke twice more before dawn, first to a cobalt sky and moonbeams lapping the bushes, heavy with moisture. In the moonlight the roots of the mapou were white, motionless, and seemingly cold. By the next time the stars had faded and light cracked the horizon. Venus had moved all the way across the sky, and now it too dimmed. I followed it until my eyes ached. A gray cloud crossed over its path, and when it was gone so was the planet. I stared and stared until I couldn't even see the sky. But it was hopeless. Venus was gone. It shouldn't have been. Astronomers know the amount of light reflected by the planet, and we should be able to see it, even in broad daylight. Some Indians can. And but a few hundred years ago, sailors from our own civilization navigated by it, following its path as easily by day as they did by night. It is simply a skill that we have lost, and I have often wondered why.

Though we frequently speak of the potential of the brain, in practice our mental capacity seems to be limited. Every human mind has the same latent capabilities, but for reasons that have always intrigued anthropologists different peoples develop it in different ways, and the distinctions, in effect, amount to unconscious cultural choices. There is a small isolated group of seminomadic Indians in the northwest Amazon whose technology is so rudimentary that until quite recently they used stone axes. Yet these same people possess a knowledge of the tropical forest that puts almost any biologist to shame. As children they learn to recognize such complex phenomena as floral pollination and fruit dispersal, to understand and accurately predict animal behavior, to anticipate the fruiting cycles of hundreds of forest trees. As adults their awareness is refined to an uncanny degree; at forty paces, for example, their hunters can smell animal urine and distinguish on the basis of scent alone which out of dozens of possible species left it. Such sensitivity is not an innate attribute of these people, any more than technological prowess is something inevitably and uniquely ours. Both are consequences of adaptive choices that resulted in the development of highly specialized but different mental skills, at the obvious expense of others. Within a culture, change also means choice. In our society, for example, we now think nothing about driving at high speeds down expressways, a task that involves countless rapid, unconscious sensory responses and decisions which, to say the least, would have intimidated our great-grandfathers. Yet in acquiring such dexterity, we have forfeited other skills like the ability to see Venus, to smell animals, to hear the weather change.

Perhaps our biggest choice came four centuries ago when we began to breed scientists. This was not something our ancestors aimed for. It was a result of historical circumstances that produced a particular way of thinking that was not necessarily better than what had come before, only different. Every society, including our own, is moved by a fundamental quest for unity; a struggle to create order out of perceived disorder, integrity in the face of diversity, consistency in the face of anomaly. This vital urge to render coherent and intelligible models of the universe is at the root of all religion, philosophy, and, of course, science. What distinguishes scientific thinking from that of traditional and, as it often turns out, nonliterate cultures is the tendency of the latter to seek the shortest possible means to achieve total understanding of their world. The vodoun society, for example, spins a web of belief that is all-inclusive, that generates an illusion of total comprehension. No matter how an outsider might view it, for the individual member of that society the illusion holds, not because of coercive force, but simply because for him there is no other way. And what's more, the belief system works; it gives meaning to the universe.

Scientific thinking is quite the opposite. We explicitly deny such comprehensive visions, and instead deliberately divide our world, our perceptions, and our confusion into however many particles are necessary to achieve understanding according to the rules of our logic. We set things apart from each other, and then what we cannot explain we dismiss with euphemisms. For example, we could ask why a tree fell over in a storm and killed a pedestrian. The scientist might suggest that the trunk was rotten and the velocity of the wind was higher than usual. But when pressed to explain why it happened at the instant when that individual passed, we would undoubtedly hear words such as chance, coincidence, and fate; terms which, in and of themselves, are quite meaningless but which conveniently leave the issue open. For the vodounist, each detail in that progression of events would have a total, immediate, and satisfactory explanation within the parameters of his belief system.

For us to doubt the conclusions of the vodounist is expected, but it is nevertheless presumptuous. For one, their system works, at least for them. What's more, for most of us our basis for accepting the models and theories of our scientists is no more solid or objective than that of the vodounist who accepts the metaphysical theology of the houngan. Few laymen know or even care to know the principles that guide science; we accept the results on faith, and like the peasant we simply defer to the accredited experts of the tradition. Yet we scientists work under the constraints of our own illusions. We assume that somehow we shall be able to divide the universe into enough infinitesimally small pieces, that somehow even according to our own rules we shall be able to comprehend these, and critically we assume that these particles, though extracted from the whole, will render meaningful conclusions about the totality. Perhaps most dangerously, we assume that in doing this, in making this kind of choice, we sacrifice nothing. But we do. I can no longer see Venus.

It was a lovely morning. The summit of the waterfall caught the first light, and as the earth moved the tips of the forest trees at the base of the ravine turned to copper. Birds spun in waves across the valley; scents too, from stoked fires, even through the mist and growing sounds, herbs and sweet woods burning in the limpid light of dawn. On all sides, like flowers, the pilgrims broke from sleep to take small drops of the sun. From the shadows I too emerged, only into a peculiar and wonderful anonymity that I had not known in Haiti up to now. In part it was my own exhaustion, but mostly it was the power of the place and the sheer exuberance of the people; for once a solitary and bedraggled *blanc* was of little interest. Asking no questions, having no past, I simply wandered, a silent witness to a sacred event unlike anything I had ever known.

All morning the trail descending to the falls quivered with the mirage of pilgrims coming and going on foot. There was no order or routine to their arrival, but it was a constant stream—as many as fifteen thousand over the course of the three-day celebration—and the basin nestled into the side of the mountain swelled like a festive carnival tent to absorb everyone. It was to be a morning of joy; one saw it on the faces of the elfin children, the young city dandies leaping over the rocks like cats, the ragged peasants laughing derisively at a fat, preposterous government official. But for the sincere it was also a moment of purification and healing, one chance each year to partake of the power of the water, to bathe and drink, and to bottle a small sample of the cold, thin blood of the divine. Already on the long, stony trail that had brought them from Ville Bonheur they had paused at least once at the tree of Legba, the guardian of the crossroads, to light a candle to invoke his support. Now before entering the water, they gathered all around the periphery of the basin where the herbalists had set up their dusty stations, displaying sooty boxes, hunks of root, loose bags of mombin leaves, and tubs of water and herbs. Houngan and mambo spoke of magic done with dew, and tied brightly colored strings to barren young women, or around the plump bellies of matrons who, in time, would dangle the strings from wax stuck to the surface of the mapou, consecrated for the blessings of the gods. In the base of one of the buttresses, a young boy hung his head as he waited to be anointed by the mambo. To one side, a houngan massaged the breast of an old woman. Scattered among the long rows of holy men and women were the hawkers pitching food for the sacrifices, tin medallions, icons, and candles. A young operator content with his gods pulled out a gameboard and dice and set up shop.

One needed only to touch the water to feel its grace, and for some it was enough to dip in the shallow silvery pools, leaving their offerings of corn and rice and cassava in small piles. But most went directly to the cascades, women and men, old and young, baring their breasts and scrambling up the wet slippery bedrock that rose in a series of steps toward the base of the highest falls. At the lip of the escarpment the river forked twice, sending not one but three waterfalls plunging over a hundred feet. What was not lost in mist struck the rocks with tremendous force, dividing again into many smaller shoots, and each one of them in turn became a sanctuary for the pilgrims. The women removed their soiled clothes, casting them to the water, and stood, arms outstretched, beseeching the spirits. Their prayers were lost to the thunderous roar of the falls, the piercing shouts, and the screams of flocks of children. Everything stood in flux. No edge and no separation—the sounds and sights, the passions, the lush soaring vegetation, primeval and rare. Young men stood directly beneath the head of the falls; the force tore off the rest of their clothes and battered their numb bodies against the rocks, but still their hands clung.

"*Ayida Wedo!*" someone called, his shout a whisper. It was true. Mist fell over the basin, and the water splintered the sunlight, leaving a rainbow arched across the entire face of the waterfall. It was the goddess of many colors, delicate and ephemeral, come to rejoice with her mate. Ayida Wedo the Rainbow and Damballah the Serpent, the father of the falling waters and the reservoir of all spiritual wisdom. Just to bathe in the cold, thin waters was to open oneself to Damballah, and already at the base of the waterfall, in the shadow of the rainbow, there were as many as a hundred pilgrims, mounted by the spirit, slithering across the wet rocks.

In the beginning, it is said, there was only the Great Serpent, whose seven thousand coils lay beneath the earth, holding it in place that it might not fall into the abysmal sea. In time, the Serpent began to move, unleashing its undulating flesh, which rose slowly into a great spiral that enveloped the Universe. In the heavens, it released stars and all the celestial bodies; on earth, it brought forth Creation, winding its way through the molten slopes to carve rivers, which like veins became the channels through which flowed the essence of all life. In the searing heat it forged metals, and rising again into the sky it cast lightning bolts to the earth that gave birth to the sacred stones. Then it lay along the path of the sun and partook of its nature.

Within its layered skin, the Serpent retained the spring of eternal life, and from the zenith it let go the waters that filled the rivers upon which the people would nurse. As the water struck the earth, the Rainbow arose, and the Serpent took her as his wife. Their love entwined them in a cosmic helix that arched across the heavens. In time their fusion gave birth to the spirit that animates blood. Women learned to filter this divine substance through their breasts to produce milk, just as men passed it through their testes to create semen. The Serpent and the Rainbow instructed women to remember these blessings once each month, and they taught men to damn the flow so that the belly might swell and bring forth new life. Then, as a final gift, they taught the people to partake of the blood as a sacrament, that they might become the spirit and embrace the wisdom of the Serpent.

For the nonbeliever, there is something profoundly disturbing about spirit possession. Its power is raw, immediate, and undeniably real, devastating in a way to those of us who do not know our gods. To witness sane and in every regard respectable individuals experiencing direct rapport with the divine fills one with either fear, which finds its natural outlet in disbelief, or deep envy. The psychologists who have attempted to understand possession from a scientific perspective have tended to fall into the former category, and perhaps because of this they have come up with some bewildering conclusions, derived in part from quite unwarranted assumptions. For one, because the mystical frame of reference of the vodounist involves issues that cannot be approached by their calculus—the existence or nonexistence of spirits, for example—the actual beliefs of the individual experiencing possession are dismissed as externalities. To the believer, the dissociation of personality that characterizes possession is the hand of divine grace; to the psychologist it is but a symptom of an "overwhelming psychic disruption." One prominent Haitian physician, acknowledging that possession occurs under strict parameters of ritual, nevertheless concluded that it was the result of "widespread pathology in the countryside which far from being the result of individual or social experience was related to the genetic character of the Haitian people," a racial psychosis, as he put it elsewhere, of a people "living on nerves." Even individuals otherwise sympathetic to vodoun have made extraordinary statements. Dr. Jean Price-Mars, one of the early intellectual champions of the religion, considered possession a behavior of "psychically disequilibrated persons with a mytho-maniacal constitution"; mythomaniacal being defined as "a conscious pathological tendency towards lying and the creation of imaginary fables."

These wordy explanations ring most hollow when they are applied to certain irrefutable physical attributes of the possessed. While recognizing, for example, the ability of the believer to place with impunity his or her hands into boiling water, another respected Haitian medical authority noted that "primitives submit coldly to surgical operations without anesthesia that would plunge us into the most terrifying shock." Rather tentatively he added that in asylums patients had been known to burn themselves and not notice even as the flesh fell away. What this well-meaning physician failed to note, among other things, was the perennial observation that the flesh of the possessed is not harmed.

What is potentially destructive in these psychological interpretations is their inherent assumption that possession is abnormal behavior, a premise that anthropologists, to their credit, have irrefutably debunked. One ethnological survey of some 488 societies around the world identified possession of some form in 360, and possession marked by trance was present in over half the total sample (including, incidentally, our own). From the Delphic oracles of ancient Greece to the shaman of northern Eurasia, possession by a spirit has been accepted as a normal phenomenon that occurs when and where appropriate, and usually within the context of religious worship. Yet even the conclusions of the anthropologists amount to observations, not explanations. They have accurately characterized the phenomenon as involving some kind of separation, transformation, and reintegration of diverse aspects of the human psyche. And, to be sure, they have been correct in noting that to some extent spirit possession is a culturally learned and reinforced response that has a therapeutic value as a spiritual catharsis.

Yet the central and disturbing questions remain. Why is it, for example, that the one possessed by the spirit in vodoun experiences total amnesia, yet still manifests the predictable and often complex behavior of the particular loa? For in Haiti, there is seldom any disagreement over the identities of the spirits. Legba is a weak old man, hobbling painfully and leaning on his crutch. Erzulie Freda is a queen, hopelessly demanding and vain. Ogoun has the warrior's passion for fire and steel, usually brandishes a machete, and often handles glowing embers. And why is it that when Ogoun does pass the flames, the one possessed is not harmed? It was upon these unanswered questions that my logic wavered. There may, in fact, be a natural explanation for these extraordinary abilities, but if so it lies in regions of consciousness and mind/ body interactions that Western psychiatry and medicine have scarcely begun to fathom. In the absence of a scientific explanation, and in the face of our own certain ignorance, it seems foolish to disregard the opinions of those who know possession best.

For most of the day, I sat near the base of the mapou. After weeks of constant travel and intermittent tension, it was good for once to remain still and simply watch, for I knew that things were about to change. Before long I was due to return to New York to report to Kline, and I didn't know when, if ever, I would come back to Haiti. Evidently Kline was satisfied with the increased samples and now was most anxious to begin further laboratory investigations. From his perspective the initial phase of the project was complete: a pharmacologically active substance had been identified, which he could promote as the material basis of the zombi phenomenon. All that remained to accomplish as far as he was concerned was the documentation and medical study of a victim as it came out of the ground.

As yet another piece of evidence, the possibility of observing a legitimate graveyard ceremony intrigued me as much as anyone, and already I had begun to make discreet inquiries on Kline's behalf. But from the start it struck me as something of a digression. The chances of success were slim, the risks were great, and moreover, even if the obvious ethical and practical difficulties could be surmounted, it would bring us no closer to what I saw as the core of the mystery. The evidence surrounding the case of Clairvius Narcisse was more than sufficient to crack the mirror of disbelief. Now for the first time the most important questions could be considered, and none of these could be answered by running all over Haiti looking for other zombis. We had the formula of the powder, and with the information provided by Herard Simon had been able to clear up a critical problem concerning the administration of the drug. From the disparate facts garnered from various informants, there was little doubt that Narcisse had suffered a form of voodoo death. But I still had not penetrated the belief system to know what the magic was, or what it had meant to Clairvius Narcisse.

The frontier of death. In Lehman's first words to me he had identified the critical issue. A zombi sits on the cusp of death, and for all peoples death is the first teacher, the first pain, the edge beyond which life as we know it ends and wonder begins. Death's essence is the severance from the mortal body of some elusive life-giving principle, and how a culture comes to understand or at least tolerate this inexorable separation to a great extent defines its mystical worldview. If zombis exist, the beliefs that mediate the phenomenon must be rooted in the very heart of the Haitian being. And to try to reach those places, to isolate the germ of the Haitian people, there was no better means than to spend a night and a day at Saut d'Eau filling my eyes with wonder, and listening to the words of the houngan.

For the Haitian, the ease with which the individual walks in and out of his spirit world is but a consequence of the remarkable dialogue that exists between man and the loa. The spirits are powerful and if offended can do great harm, but they are also predictable and if propitiated will gratefully provide all the benefits of health, fertility, and prosperity. But just as man must honor the spirits, so the loa are dependent on man, for the human body is their receptacle. Usually they arrive during a religious ceremony, ascending up the axis of the poteau mitan, called forth by the rhythm of the drums or the vibration of a bell. Once mounted, the person loses all consciousness and sense of self; he or she becomes the spirit, taking on its persona and powers. That, of course, is why the body of the possessed cannot be harmed.

But the human form is by no means just an empty vessel for the gods. Rather it is the critical and single locus where a number of sacred forces may converge, and within the overall vodoun quest for unity it is the fulcrum upon which harmony and balance may be finally achieved. The players in this drama are the basic components of man: the *z'étoile*, the *gros bon ange*, the *ti bon ange*, the *n'âme*, and the *corps cadavre*. The latter is the body itself, the flesh and the blood. The n'âme is the spirit of the flesh that allows each cell of the body to function. It is the residual presence of the n'âme, for example, that gives form to the corpse long after the clinical death of the body. The n'âme is a gift from God, which upon the death of the corps cadavre begins to pass slowly into the organisms of the soil; the gradual decomposition of the corpse is the result of this slow transferral of energy, a process that takes eighteen months to complete. Because of this, no coffin may be disturbed until it has been in the ground for that period of time.

The z'étoile is the one spiritual component that resides not in the body but in the sky. It is the individual's star of destiny, and is viewed as a calabash that carries one's hope and all the many ordered events for the next life of the soul, a blueprint that will be a function of the course of the previous lifetime. If the shooting star is bright, so shall be the future of the individual.

The two aspects of the vodoun soul, the ti bon ange and the gros bon ange, are best explained by a metaphor commonly used by the Haitians themselves. Sometimes when one stands in the late afternoon light the body casts a double shadow, a dark core and then a lighter penumbra, faint like the halo that sometimes surrounds the full moon. This ephemeral fringe is the ti bon ange—the "little good angel"—while the image at the center is the gros bon ange, the "big good angel." The latter is the life force that all sentient beings share; it enters the individual at conception and functions only to keep the body alive. At clinical death, it returns immediately to God and once again becomes part of the great reservoir of energy that supports all life. But if the gros bon ange is undifferentiated energy, the ti bon ange is that part of the soul directly associated with the individual. As the gros bon ange provides each person with the power to act, it is the ti bon ange that molds the individual sentiments within each act. It is one's aura, and the source of all personality, character, and willpower.

As the essence of one's individuality, the ti bon ange is the logical target of sorcery, a danger that is compounded by the ease and frequency with which it dissociates from the body. It is the ti bon ange, for example, that travels during sleep to experience dreams. Similarly, the brief sensation of emptiness that immediately follows a sudden scare is due to its temporary flight. And predictably it is the ti bon ange that is displaced during possession when the believer takes on the persona of the loa.

At the same time, because it is the ti bon ange that experiences life, it represents a precious accumulation of knowledge that must not be squandered or lost. If and only if it is protected from sorcery and permitted to complete its proper cycle, the ti bon ange may be salvaged upon the death of the individual and its legacy preserved. Only in this way may the wisdom of past lives be marshaled to serve the pressing needs of the living. A great deal of ritual effort, therefore, must be expended to secure its safe and effortless metamorphosis. At initiation, for example, the ti bon ange may be extracted from the body and housed in a *canari*, a clay jar that is placed in the inner sanctum of the hounfour. In this way the ti bon ange may continue to animate the living while remaining directly within the protective custody of the houngan. Yet even here there are no guarantees. Though it is difficult to kill the one whose ti bon ange has been placed in a canari, if the magic used against the individual is strong enough, the resulting misery may be so great that he will ask the houngan to release the soul that he might end his ordeal. And even if the individual does survive life, he is still at risk in death, for with the demise of the corps cadavre the houngan must break the canari so that the ti bon ange may return to hover about the body for seven days. Then, since the vodounist does not believe in the physical resurrection of the body, the soul must be definitively separated from the flesh, and this takes place during the *Dessounin*, which is the major death ritual. Throughout this period the ti bon ange is extremely vulnerable, and it is not until it is liberated from the flesh to descend below the dark abysmal waters that it is relatively safe.

The ti bon ange remains below in the world of *Les Invisibles* for one day and one year and then, in one of the most important of vodoun ceremonies—the *Wété Mo Nan Dlo*—it is reclaimed by the living and given new form. In place of the body that has decayed, the soul, now regarded as an "esprit," is deposited in another clay jar called a *govi*. To the Haitian this reclamation of the dead is not an isolated sentimental act; on the contrary, it is considered as fundamental and inescapable as birth itself. One emerges from the womb an animal, the spiritual birth at initiation makes one human, but it is this final reemergence that marks one's birth as divine essence. The spirits in the govi are fed and clothed and then released to the forest to dwell in trees and grottos, where they wait to be reborn. After the last of sixteen incarnations, the esprit goes to Damballah Wedo, where it becomes undifferentiated as part of the *Djo*, the cosmic breath that envelops the universe.

This lengthy passage of the ti bon ange corresponds to the metamorphosis of the individual human into pure spiritual energy. Hence, with the successive passing of generations, the individual identified with the esprit in the govi is transformed from the ancestor of a particular lineage into the generalized ancestor of all mankind. Yet even this pure spiritual energy must be made to serve, and for it to function it must become manifest. Thus from the ancestral pool there emerge archetypes, and these are the loa. It is, of course, possession, the return of the spirits to the body of man, that completes the sacred cycle: from man to ancestor, ancestor to cosmic principle, principle to personage, and personage returning to displace the identity of man. Hence, while the vodounist serves his gods, he also gives birth to them, and this is something that is never forgotten; as much as the spirit is the source of the flesh, so the flesh gives rise to the spirit. In place of opposition between the two, there is mutual dependence. Thus the regular arrival of the divine is not considered miraculous, but rather inevitable.

Within this cosmic exchange, perhaps man's most critical contribution is the preservation of his own equilibrium, for without it the receptacle of the gods is placed in danger. The ideal form of man, therefore, is one of coherence, wherein all the sacred components of the individual find their proper place. The maintenance or restoration of this balance is the duty of the houngan, and it accounts for his unique role as healer. In our secular society, life and death are defined in strictly clinical terms by physicians, with the fate of the spirit being relegated to the domain of religious specialists who, significantly, have nothing to say about the physical well-being of the living. In vodoun society, the physician is also the priest, for the condition of the spirit is as important as—and in fact, determines—the physical state of the body. Good or bad health results not from the presence or absence of pathogens but from the proper or improper balance of the individual. Sickness is disruption, imbalance, and the manifestation of malevolent forces in the flesh. Health is a state of harmony, and for the vodounist it is something holy, like a perfect service for the gods.

As a result, vodoun medicine acts on two quite different levels. There is an entire range of relatively minor ailments that are treated symptomatically much as we would, only with medicinal plants and folk preparations, many of which are pharmacologically active. A basic knowledge of the leaves in such profane treatments is part of the traditional education of virtually every rural Haitian, and though there are respected specialists known as *dokte feuilles*—leaf doctors—their expertise is considered mundane. Much more serious are the troubles that arise when the harmony of the spiritual components is broken. Here it is the source of the disorder, not its particular manifestation, that must be treated, and that responsibility falls strictly within the domain of the houngan. Since disharmony will affect all aspects of the individual's life, problems brought to the houngan include both psychological and physical ailments, as well as other troubles such as chronic bad luck, marital difficulties, or financial problems. Each case is treated as unique. As a form of medicine, it does not ignore the existence of pathogens, it simply comments that the pathogens are present in the environment at all times, and it asks why certain individuals succumb when they do.

To restore the patient's health may involve a number of techniques. At the material level these include herbal baths and massage, physical isolation of the patient in the hounfour, administration of medicinal plant potions, and perhaps most importantly, a sacrifice, that the patient may return to the earth a gift of life's vital energy. But it is intervention on the spiritual plane that ultimately determines the patient's fate, and for this the houngan is but a servant of the loa. The spirit is called into the head of either the houngan or an assistant, and like an oracle the physical body of man dispenses the knowledge of the gods.

Inevitably, there are times when the forces arrayed against the individual are simply too powerful. If disharmony at the core of man results in sickness, the irrevocable separation of the spiritual components will bring death. But death, like life, stretches far beyond the temporal limits of the body. Life begins not at physical conception but at an earlier moment when God first decides that a person should exist. Death is not defined just by the passing of the flesh, but as the moment when all the spiritual components find their proper destination. Thus the vodoun adept, believing in the immortality of the spirit, fears death not for its finality but because it is a critical and dangerous passage during which the five vital aspects of man dissociate, leaving the ti bon ange, in particular, vulnerable to capture by the sorcerer.

But the death of the body brings other equally pressing concerns, for there are two possible causes of death, and the implications are profoundly different. Rarely, as in the case of an old man passing away in his sleep, a death may be considered natural, a call from God (*mort bon Dieu*) and beyond the influence of man. Unnatural deaths include all those we might label as "before one's time," and more often than not these result from the intervention of sorcery. And by vodoun definition, anyone who suffers an unnatural magical death may be raised as a zombi. At times it may be in the interest of the bokor to cause the unnatural death, and there are countless ways of doing so. But causing the unnatural death does not create a zombi; it just makes the victim immediately susceptible. Once this is understood, it becomes apparent that our entire investigation was based on an unwarranted assumption.

Since we knew that the zombi powders could pharmacologically induce a state of apparent death, we had all assumed that the bokors recognized an explicit cause-effect relationship between the powder and the resurrection from the grave. Clearly, I now realized, the vodounist did not necessarily consider it such a linear process. For them the creation of a zombi involved two only indirectly related events: the unnatural death and the graveyard ceremony. According to their beliefs, the powders just kill, and as in the case of any unnatural death the victim of the powder may be raised as a zombi. It is not the antidote or a powder that creates a zombi; it is the magical force of the bokor. That was why Herard Simon could rest assured that I would never understand zombis. I did not know the magic, and he himself did not know, nor did he believe that anyone would share those mysteries with me.

For the vodounist, then, zombis are created by sorcery, and it is the belief in the magic that makes the relatives of the dead concerned. For good reasons, they go to great efforts to ensure that the dead are truly dead, or at least protected from such a horrible fate. This is why the body may be killed again, with a knife through the heart or by decapitation. And this explains why seeds may be placed in the coffin so that whoever appears to take the body will be obliged to count them, a task that will take him perilously into the dawn.

To create a zombi, the bokor must capture the ti bon ange of the intended victim, a magical act that may be accomplished in a variety of ways. A particularly powerful bokor, for example, may through his spells gain control of the ti bon ange of a sailor who dies at sea or of a Haitian who is killed in a foreign land. Alternatively, the bokor may capture the ti bon ange of the living and hence indirectly cause the unnatural death: the individual, left without intelligence or will, slowly perishes. One way of thus capturing the ti bon ange is to spread poisons in the form of a cross on the threshold of the victim's doorway. The magical skill of the bokor guarantees that only the victim will suffer. This, of course, was the service that LaBonté and Obin had offered me at Petite Rivière de Nippes. Yet a third means of gaining control of the ti bon ange is to capture it immediately following the death of the corps cadavre during the seven days that it hovers around the corpse. Hence the bokor may or may not be responsible for the unnatural death of the victim, and the ti bon ange may be captured by magic before or after the death of the corps cadavre.

Whatever the circumstances, the capture of the ti bon ange effects a split in the spiritual components of the victim and creates not one but two complementary kinds of zombis. That is what Herard had shown me. The spirit zombi, or the zombi of the ti bon ange alone, is carefully stored in a jar and may later be magically transmuted into insects, animals, or humans in order to accomplish the particular work of the bokor. The remaining spiritual components of man, the n'âme, the gros bon ange, and the z'étoile, together form the zombi cadavre, the zombi of the flesh.

Now the resurrection of the zombi cadavre in the graveyard requires a particularly sophisticated knowledge of magic. Above all, the bokor must prevent the transformations of the various spiritual components that would normally occur at the death of the body. First the ti bon ange—which may float above the body like a phosphorescent shadow—must be captured and prevented from reentering the victim. One way to assure this is to beat the victim violently, as occurred with Narcisse. Secondly, the gros bon ange must be prevented from returning to its source. Thirdly, the n'âme must be retained to keep the flesh from decaying. The zombi cadavre with its gros bon ange and n'âme can function; however, separated from the ti bon ange, the body is but an empty vessel, subject to the direction of the bokor or whoever maintains control of the zombi ti bon ange, Herard's zombi astral. It is the notion of alien, malevolent forces thus taking control of the individual that is so terrifying to the vodounist. In Haiti, the fear is not of being harmed by zombis; it is fear of becoming one. The zombi cadavre, then, is a body without a complete soul, matter without morality.

In the end, the solution to this aspect of the zombi mystery had a certain elegance. For the vodounist the creation of a zombi is essentially a magical process. However, the bokor in creating a zombi cadavre may cause the prerequisite unnatural death not by capturing the ti bon ange of the living but by means of a slow-acting poison that is applied directly to the intended victim. Rubbed into a wound or inhaled, the poison kills the corps cadavre slowly, efficiently, and discreetly. That poison contains tetrodotoxin, which acts to lower dramatically the metabolic rate of the victim almost to the point of clinical death. Pronounced dead by attending physicians, and considered materially dead by family members and even by the bokor himself, the victim is in fact buried alive. Undoubtedly in many instances the victim does die either from the poison itself or by suffocation in the coffin. The widespread belief in the reality of zombis in Haiti, however, is based on those cases in which the victim receives the correct dose of the poison, wakes up in the coffin, and is taken from the grave by the bokor. The victim, affected by the drug, traumatized by the set and setting of the total experience, is bound and led before a cross to be baptized with a new name. After the baptism, or sometimes the next day, he or she is made to eat a paste containing a strong dose of a potent psychoactive drug, the zombi's cucumber, which brings on a state of disorientation and amnesia. During the course of that intoxication, the zombi is taken away into the night.

There remained one haunting question. If the formula of the poison and the sorcerer's spell explained how a zombi succumbed, it said nothing about why he was chosen. Many, including those who had formulated Haiti's national laws, had concluded that zombification was a random criminal activity, yet another symptom of a conspiracy of fear that was presumed to be the common plight of the peasant. But the longer I remained in Haiti, and the more I learned of the vodoun society, the more impressed I was by its internal cohesion. Sorcery was certainly a potent force to be dealt with, but to a great extent it had been institutionalized as a critical component of the worldview. To ask why there is sorcery in Haiti is to ask why there is evil in the world, and the answer, if there is one, is the same as that provided by all the great religions: evil is the mirror of good, the necessary complement that completes the whole of creation. The Haitians as much as any people are conscious of this sacred balance.

So the suggestion that zombification was a haphazard phenomenon ran in the face of my data. In obtaining the various preparations, I had come into direct contact with a number of secret societies, and in certain instances it had been their leaders who controlled the powder. There was strong circumstantial evidence that both Narcisse and Ti Femme had been exceedingly unpopular in their respective communities, and I had it from a number of sources that zombification was a process that included some kind of judgment before a tribunal. Max Beauvoir had gone so far as to suggest that the answer to the mystery lay within the councils of the secret societies. But what did this mean? In my search for the poison, instinct unfettered by bias had served me well, but as I probed deeper my intuitions were increasingly clouded by my ignorance. The secret societies—who were these groups, what was the nature of their organization, and how did they relate to the other national authorities? It was with the desire to explore these questions, as much as on account of Kline's summons, that I returned to America. I would have to go back in time to the beginning and the harsh days of the French colony.

11

Tell My Horse

IT HAPPENED on a plantation near Limbé in the year 1740. At first even the man himself did not notice the iron rollers of the cane press flush crimson with his own blood. By the time the child's scream alerted the driver to slice the leather traces connecting the horse to the shaft of the mill, the arm was crushed to the shoulder, and the blood mixed freely with the sweet sap of the cane. Pain was not new to the slave, and what he felt now was numbed by the rage of an intolerable impotence. His free hand flailed at the press, and with all the force of his sinuous body he pulled back, reversing the rollers, withdrawing fragments of his mangled arm. Delirium took him, leading him back on a hallucinatory passage to the land of his birth, to the Kingdoms of Fula and Mandingo, to the great cities of Guinea, the fortresses and vast markets that drew traders from an entire continent and beyond, the temples that made a mockery of the paltry buildings in which the French worshipped their feeble god. He never noticed the rope tourniquet placed around his shoulder to seal the flow of blood, nor did he hear the call for the machete that would complete the crude work of the press. He felt only the beginnings of the sound, of a single syllable rising from the base of his bloodstained legs, recoiling through the hollow of his gut until what left his lips was no longer his. It was the wrathful call of crystallized hatred, a cry of vengeance, not for himself, but for an entire people stolen from Africa and dragged in chains to the Americas to work land stolen from the Indians.

François Macandal should have died, but the Mandingue slave was no ordinary man. Even before the accident he was a leader among the slaves of the northern district around Limbé. By day, they had watched him endure the cruelties of the overseers with indifference, his bloodshot eyes casting scorn at the whips of knotted cord, or the stretched and dried penis of a bull. By night he had calmed the people with his eloquence, spinning tales of Guinée that had emboldened even the most dispirited of men. When he spoke people considered it an honor to sit by his side, and as he slept the women vied for the chance to share his bed, for his dreams were revelations that allowed him and those by him to see into the future. But it was the fearless way Macandal endured the accident in the mill that confirmed what the people had always suspected. Only the whites could fail to note that Macandal was immortal, an envoy of the gods who would never be vanquished.

The accident freed Macandal to wander. No longer fit to work in the fields, he was made a herder and sent out each daybreak to drive the cattle into their mountain pastures. No one knew what he did during the long hours away from the plantation. Some said he discovered the magic in plants, foraging for leaves that mimicked the herbs he had known in Africa. Others said he sought out the old masters who dwelled in caves, whose footsteps alone caused the earth to tremble. Only one thing was certain. Macandal in his wanderings was not alone, for the mountains around Limbé were one of the refuges of the thousands of Africans who had fled the plantations, runaway slaves with a price on their heads and known to the French as Maroons.

Bloated by wealth unlike anything seen since the early days of the conquest, the colonial planters of Saint Domingue made an institution of cruelty. Field hands caught eating cane were forced to wear tin muzzles while they worked. Runaways had their hamstrings sliced. Brandings, indiscriminate floggings, rape, and killings were a matter of course, and for the slightest infraction a man was hung from a nail driven through his ear. Slaves like cane were grist for the mill, and the death toll in some years rose as high as eighteen thousand. The documented excesses of certain owners almost defy belief. One slave was kept in chains for twenty-five years. A notorious planter was known always to carry a hammer and nails just to be prepared to hang from the trees the severed ears of those he punished. Other common tortures included spraying the flesh with boiling cane syrup, sewing the lips together with brass wire, castration and sexual mutilation of both men and women, live burial, binding men—their skin glazed with molasses—across the paths of ants, enclosing people in barrels with inward protruding nails, and stuffing the anus with gunpowder which was then ignited, a practice common enough to give rise to the colloquial expression "blasting a black's ass." So systemic was the abuse of the slaves that it supported a profession of executioners whose fees were regulated by law. The charge to burn a man alive, for example, was set at sixty French pounds. A hanging was only thirty, and for a mere five pounds you could have a slave branded and his ears cut off.

Such savagery was the rule, not the exception, and the fact that first the Indians and then thousands of indentured whites—petty thieves, convicts, or simply urban poor kidnapped in the port cities of Europe— had toiled in servitude before them did little to still the rage of the Africans. Forced labor was the foundation of an economic system that knew no color boundaries; like an open sore the plantations grew upon the Caribbean, and when the Indians died, and the supply of white trash failed to meet demand, the merchants tapped deeper into Africa, drawing away men and women not because they were black, but because they were cheap, limitless in number, and better. European class societies whose elite thought nothing of hanging an English child for petty theft, or packing indentured workers, white or black, like herring into the festering holds of ships, cared less about the origins of their laborers than the production of their labor. Slavery was not born of racism; rather, racism was the consequence of slavery. In the first days of colonialism, when the merchants sailed away from the insular world of Europe, the color of the worker's skin meant no more to them than it did to the kings of Africa, rulers who lorded over thousands of their own slaves, and who for a suitable profit were more than willing to pass them along. Of course, all of this mattered little to the men and women unloaded into bondage in Saint Domingue. For them the enemy had a face, and there was no doubt as to its color.

Confronted by unrelenting intimidation and torture, the options of the slaves were few. Those who chose to submit and endure did what was necessary to ease their plight: self-inflicted wounds allowed a respite from the brutal labor in the fields, women gave themselves to the overseers, or spared their unborn children by practicing abortion. Others sought immediate relief in suicide. But those who could not be broken and whose desire for freedom drove them to accept desperate measures fled the plantations under the cover of night. Some remained close by, hiding out by day and dependent on the collusion of their families and friends that remained behind. Others, especially the skilled workers fluent in Creole, slipped into anonymity in the cities, passing themselves off as freedmen and seeking work among the faceless crowds in the markets or on the docks. Still others made for the Spanish frontier of Santo Domingo, overland across the savannas and mountains and through the dense forests that still blanketed the island. Like stray cattle, these runaways were but a nuisance to the planters, readily dealt with by professional bounty hunters and their dogs. If recaptured on the fringe of the plantation, they were simply returned to be flogged and publicly humiliated, made to kneel outside the white man's church to beg forgiveness for "insubordination to the situation in which God had placed him." Should the runaway be unfortunate enough to be discovered some distance away, or should he resist capture, he was summarily killed—shot and ravaged by the hounds—once his identifying brand had been sliced away from his skin.

But there was another type of Maroon, men like Macandal who were not content to hover in the shadows like animals, or waste away in the limestone sinkholes and caves that dotted the land. These were Africans who would take responsibility for their fate, men who sought not just to survive but to fight and to seek revenge for the weight of injustice that had tormented their people. When these men and women left the plantations, taking with them anything of value they had managed to pilfer—a mule, knife, machete, field tools, clothing—they joined the organized bands in remote sanctuaries deep in the hinterland. There they lived in armed camps, sealed off by palisades surrounded by wide ditches, fortified at the bottom by pointed stakes. They cleared gardens, and to a great extent were self-sufficient, supplementing what they grew with periodic raids on the plantations. If solitary runaways were a mere irritation to the French, these independent Maroon retreats were no less than training grounds for guerrilla fighters that threatened the order and stability of the entire colony.

The French regime responded by waging an incessant campaign

of extermination. Specialized military forces known as *marcehaussée* were maintained and sent on frequent and costly forays into the mountains. Some of these expeditions were moderately successful, returning with captives who were publicly broken on the wheel. Others never came back at all. And not one was able to penetrate or destroy the principal strongholds. For the French could not be everywhere, and the Maroons were—in the mountains rising behind the plantations of the northern plain at Cap Francis, the Cul-de-Sac near Port-au-Prince or the rolling valleys near Cayes in the south. As a result, by the mid-eighteenth century entire regions were effectively sealed off to whites. One rebellion that covered a vast mountain block in the south lasted a hundred years, until the French finally abandoned the zone altogether. Further north a Maroon community in the Bahoruco Mountains thrived for eighty-five years, until the French proposed a truce under the terms of which the Maroons would be permitted to form an independent clan. When the leader of that particular band of rebels arrived to negotiate, it was discovered that he had been a Maroon for over forty-five years.

As the French military expeditions collapsed in the mountains, the colonial administration did what it could to destroy the clandestine network that maintained the flow of goods and information between the plantations and the Maroons. Fear of the rebels was behind the constant legislation restricting the movement and normal interaction between the slaves. Blacks were prohibited from going out at night, visiting neighboring plantations, using boats, or even talking among themselves without the master's permission. Night searches were frequent, and anyone caught with weapons or aiding runaways was brutally and publicly punished. But at a time when slaves outnumbered whites a hundred to one on the plantations, there was really very little the French could do. For even if some of the slaves came to fear the wrath and disruption of the Maroons as much as they did the whip of the planters, there could be no doubt that the rebel bands fought for freedom, and as a result with each successive generation their legend grew. With increasing impunity, the guerrillas came out of the hills, raiding stores, pillaging plantations, and all the while spreading along with terror the idea of liberty. By 1770, according to a contemporary report, the number of Maroons had increased to such proportions that "security became nonexistent" and it was unwise to wander alone in the hills.

Just who and how many chose to follow this desperate path is uncertain, but colonial records provide some clues. Between the years 1764 and 1793, for example, newspaper advertisements alone indicate some forty-eight thousand cases of Maroonage. How many of these ended up in the Maroon enclaves is not known, but the figure does provide a sense of the scale of the problem that faced the French. Significantly, a large percentage of those who did flee had not lived in the colony more than a year, and many escaped virtually off the docks. One colonial document covering a single port for a fifteen-day period in January of 1786 lists 43 new slaves escaped or recaptured. In 1788, out of 10,573 slaves disembarked over a ten-month period at Cap Francis, more than 2,000 got away. Critically, while the Creole Maroon could slip inconspicuously into the bowels of the city, these fresh arrivals from Africa, ignorant of the ways of the colony, were the ones invariably to flee to the hills. Thus a good many of the recruits to the Maroon communities were the individuals least socialized into the regime of the whites. Into their new homes, then, they brought not the burdens of slavery but the ways of Africa.

Behind a veil of secrecy that alone allowed them to survive, these Maroon communities developed genuine political, economic, and religious systems of their own. Their leaders were culled from what contemporary observers described as a "new class of slave" that arrived in the colony throughout the eighteenth century. These were men of royal blood, often educated not just by their own oral traditions but by Arab teachers, and endowed by birth with intelligence, moral vigor, and the call of a militant tradition. Their people were also a chosen lot, mostly young men between the ages of seventeen and thirty-five, each one the end product of a tortuous selective process. Merely to reach the Maroon camps implied surviving the brutal passage from Africa, enduring the abuse on the plantations, and then outwitting the hounds of the bounty hunters to face willingly a life of daily risk, physical deprivation, and constant adversity.

Acceptance into the ranks of the Maroons was strictly controlled. Only those who came voluntarily were taken, and these only after making sure that they were not colonial spies. Blacks captured during raids could be made slaves, and on the slightest suggestion of betrayal were put to death. Newly arrived runaways had first to erase their past, mutilating their brands with knives or the juice of toxic plants—*acajou* or bresillet—that caused disfiguring welts. They endured rigorous initiations in which they learned the secret handshakes and passwords that would distinguish friend from foe during the raids. Publicly they swore allegiance to the community, and discovered in graphic terms what would occur should they betray the secrets of the group. If secrecy defined and protected the integrity of these communities, the obvious models for their internal organization were the secret societies that at least some of the ex-slaves must have joined in their youth in Africa.

As they are today, during the colonial era secret societies were a dominant social force throughout much of West Africa, particularly among the coastal rain forest peoples who were taken in bondage to Saint Domingue. The parallels between these groups and what later evolved in the Maroon communities in the colony are striking. Membership was by initiation, a lengthy process that exposed the candidate to physical hardships, tests of endurance and pain, following which he learned the secret passwords, symbols, and handshakes of the society. As in Saint Domingue, it was knowledge of these esoteric signs that defined the group; in virtually every other regard the societies were not secret, and in fact their function demanded that their existence be completely known. For these societies were no mere peripheral feature of West African culture; they lay at the very core and remained, both before and after the colonial era, the principal and militant champion of the traditional way of life. The Poro society of Sierra Leone, for example, left its mark on virtually every facet of Mende life, taking responsibility for tribal education, the regulation of sexual conduct, the supervision of political and economic affairs, as well as the operation of social services including recreation and medical care. A key to the strength of the West African secret societies—and what was particularly applicable to the needs of the Maroons—was the fact that their interests and activities were defined in terms of the community, rather than a lineage or clan. Thus they provided for the Maroon bands an invaluable model for the consolidation of the diverse cultural backgrounds of the individual slaves.

As the single most important arbiter of culture, the West African secret societies had as a vital function the administration of justice, and as in the case of the leopard society among the Efik of Old Calabar, their tribunals delivered a verdict based on the outcome of the poison ordeal. Judgment by ordeal could cover any and all personal or social crimes, and was inevitably invoked in suspected cases of sorcery. Not surprisingly, the secret societies developed a particularly refined knowledge of toxic preparations, learning not only to identify and experiment with different species of plant and animal, but also to control dosage, means of administration, and even the psychological set of the potential recipient.

But the use of toxic preparations was not restricted to the secret societies, nor even among them to the ordeal tribunals. Perhaps as much as any single material trait, the manipulation of poisons remains a consistent theme through African cultures. In certain regions, for example, criminals were executed by pricking their skin with lances or needles dipped in the juice of toxic plants. In parts of West Africa when a king died, his heir had to submit at least twice to poison ordeals to prove his supernatural strength; should he fail and die, the lineage was broken and the throne was declared vacant. Individual sorcerers often used potions of course, but in one of the most extraordinary developments, poisons were used systematically by established rulers in vain atempts to purge entire populations of evil. The leaders of the Cazamance and Balantes peoples in West Africa, for example, used a preparation based in part on the bark of a tree known as *tali* (*Erythrophleum guineense*—Leguminosae). Other ingredients included the powder ground from the dried hearts of previous victims and a number of admixtures reminiscent of those used in contemporary Haiti—ground glass, lizards, toads, crushed snakes, and human remains. Placed in a vat and allowed to ferment for a year, this toxic preparation was then ceremoniously paraded on a day of great festivities and given to every citizen. Each year as many as two thousand people died. With the advent of the slave trade, however, West African rulers discovered an even better means of purging their societies. According to Moreau de Saint-Mery, the most respected of the early colonial authorities, certain kings made use of their treaties with the European merchants to get rid of their poisoners by condemning them to deportation to Saint Domingue.

When Macandal finally fled the plantation he buried his vengeance deep within his breast. He didn't fear recapture; no planter would risk the price of valuable hounds to run down a one-armed slave, let alone a Mandingue whose very tribal name was synonymous with rebellion. Macandal's concern was one of timing—when and how to best seek retribution. It would be easy enough to kill. Muskets and powder were available for a price from the freedmen in the cities, and even armed with clubs and a few machetes a dozen Maroons could handily overrun a plantation, burning the stables, the drying sheds, and the cursed mills, and still have time to pay their respects to the master's wife before the militia arrived. But pillaging plantations and sacking mansions now held little attraction for Macandal, and for what he intended to do there weren't enough guns in all the colony. Once when asked to interpret one of his dreams, he had called for a clay vessel and placed into it three kerchiefs. He pulled out one that was yellow and explained to the crowd that it was the color of the ones who were born on the land. The next one was white, and it was they who now ruled. And here, he said, finally, are the ones that shall remain masters of the island, and the color of the kerchief was black.

For six years the network spread, with the whites completely oblivious of the danger about to befall them. Macandal was everywhere, moving with the impunity of a god garnering favors and awakening the zeal of the people. He chose his agents with care and placed them in every corner of the colony, and they in turn kept him informed of potential converts—perhaps a coachman whose woman had been recently raped by the master, or a kitchen boy whipped raw for stealing bread. On parchment and bark he drew figures in charcoal, tallying in symbols that only he understood the names of the plantations and the leaders of each and every work gang that toiled beneath the relentless sun—every man and woman who would come to his cause. He wanted everyone; they could look on him as they chose so long as their eyes held terror; be it fear of God or man it mattered not to him, just terror so that the secret would remain safe. But above all he needed those who worked on the inside—the coachmen, cooks, and domestic servants, the ones whose presence in the very bosom of the whites would not sound the alarm. In that way, Macandal would be assured that with each anguished dying breath the slavers would stare into the faces of those closest to them and see only the reflection of their own evil smiles.

By night Macandal wandered, but by day he ran a school, the students spread out before him in the thick grass, their fingers releasing the musty odor of fungi, the ooze of molds, and the pungent scent of crushed venom glands. The men held Macandal in awe, and at his behest combed the island, bringing back herbs with sap that stung, evil-looking sea creatures, snakes, and toads. Together they reached deep into their collective memory, struggling with senses numb from disuse to remember the lessons taught in their youth, the formulae and preparations and ingredients that might be mirrored among the plants and animals of the new land. If Macandal was their teacher, he in turn was the apprentice of others, the old women and men who lived alone among the dripping stalactites in caves and slept on beds of bat droppings. It was they who retained the ancient wisdom, some learned in Africa and some acquired on the island from the descendants of those who had lived among the remnants of the Arawakans, the tortured sons of the *caciques* that had ruled the land before the arrival of the whites. In the caves, these elders studied Macandal's discoveries, fingering the fruits that bled red and blackened in the air, the shriveled lizards, and venomous insects. What they finally approved, after years of study, Macandal placed in the belly of the mortar, broken at the edges and worn with use. Then before their eyes he ground the silent death that would one day walk across the fields, and reach into every kitchen in the land.

In time, like ink on a blotter, the poison seeped into the lives of the whites. First the cattle died, one by one, until the stiff carcasses littered the northern plain. The planters in dismay hired the best scientific minds, herbalists who left their physic gardens in the Cap to tramp across the pastures in search of some vile weed fouling the fields. For days and weeks they combed the grass. Just as they thought that they had found the guilty plant, and the work gangs had begun to sweep clean the pastures, the first of the dogs died, and word came that poison had entered the houses.

To their horror, the whites found themselves in a trap of their own making, dependent on the very people who were the agents of their doom. The poison appeared everywhere: baked into bread, in medicine vials, in kegs of ale lifted directly from the ships and drunk because the water from the wells could no longer be trusted. Entire banquets succumbed, sometimes from the soup, perhaps the tea, the wine, or even fruit picked fresh from the trees. The terror of the whites gave way to rage, and innocent slaves were flayed alive. The slightest suspicion of collaboration with the poisoners meant a horrible death. But the enemy could not be seen; only its mark was felt, universally on the whites and equally on any black who showed signs of betraying the agents of Macandal. The colonial administration declared a state of siege and emptied the garrisons to parade up and down the streets of the Cap, their guns shouldered and useless against the invisible enemy. The courts condemned whoever they imagined to be guilty, and work gangs were decimated in attempts to secure the names of the leaders of the conspiracy. The chemists and herbalists reconvened to attempt once more to identify the source of the plague, whether animal or plant, or perhaps some compound taken from the apothecary or some potion brought by the wretches from Africa. A royal proclamation prohibited any slave from concocting any remedy, or attempting to cure any sickness with the exception of snakebite. But nothing that the government did could stop the contagion. Hundreds of slaves died, and as many whites. Before Macandal was through, six thousand at least would be dead.

It was a child that finally betrayed him, a young girl arrested along with three men as poisoners and condemned to be burned alive. One by one she was made to suffer the agonies of the others, watching the flames grow from the base of the pyre until they flared out of the acrid smoke to ignite the creosote-soaked rags. The smell of the flesh turned her stomach, and as she tried to look away they twisted her face to the sight and held it until she saw the belly of the men bloat, bubbling at the surface, dripping fluid until the skin, stretched to the limit, split, disgorging the steaming entrails of the gut. When her turn came, the executioner tormented her with the pinewood torch, tracing patterns in the air, brushing its burning end close to her skin. Her horror grew with her rage and fear, and just as the order was given, she broke, releasing the names of fifty others, which were dutifully recorded before the executioner, ignoring her cry of protest, went ahead and dropped the burning ember onto the base of the pyre.

The web of betrayal grew until it enveloped Macandal himself. But when his turn came, and he was paraded stripped to the waist before the people assembled in the capital, the citizens in their silk waistcoats beneath festive parasols, the slaves standing ebony black and solemn, the soldiers cautious, moving at the pace of death with the rhythm tapped out on drums covered in dark cloth, a strange thing happened. Macandal looked neither frightened nor even defiant as they lashed him to the post and brought forth the torch; he seemed indifferent, almost bored as he waited for the event, so carefully orchestrated by the governor, to be over. And when they saw him like that, a murmur ran through the ranks of the slaves, and their normally inscrutable expressions became radiant, expectant, disquieting to those whites sensitive enough to notice. Then, as the first flames reached the base of his legs, Macandal lifted his face, screaming at the sun. His body began to shake violently, his torso thrusting away from the post, his free stump pounding the air until with a single spasm that drew the breath out of the crowd he broke free and flung himself beyond the flames. Pandemonium broke over the mob. Amid calls of "Macandal is free!" the whites fled the plaza, and the guards rushed to the pyre, claiming later that they had recaptured the slave and bound him to a plank and cast him back onto the flames. But none of the blacks saw it done, and though the governor even produced the ashes to quell the fears of the whites, it was to little effect. The entire northern province, lulled into complacency by his capture, rang out with the alarm, and once again the planters felt like prisoners barricaded within their own houses. For the blacks there was little doubt as to what had occurred. If captured, Macandal had always told them, he would turn himself into a fly, and no one questioned his ability to do so, especially when after his reputed death, the toll of poison continued as before.

Macandal's was not the first, nor certainly the last, Maroon revolt to shake the foundations of the colony. As early as 1681, before the colony had passed from the Spanish to the French and at a time when there were as many indentured whites as Africans, with a total population of only about six thousand, Maroonage was already an acknowledged threat. Two years before, in one of the earliest documented revolts, a slave named Padrejean had killed his master, recruited a band of twenty Africans, and embarked on his goal of strangling every white in the land. The revolt failed, but it was the type of incident that drew the attention of the king and led to the royal edict of 1685, a law that among other injunctions specified that a captured Maroon have his ears cut off and a shoulder branded with a fleur-de-lys; should the offense be repeated the hamstrings would be cut and a second brand applied to the other shoulder. The publication of this decree was an indication of a growing concern among the free whites, a fear that would become hysteria as the population of slaves soared. By the early years of the eighteenth century seditious plots, mysterious killings, and rumors of impending catastrophe became a staple of colonial life. Poisons were already so common that in 1738, two years before Macandal even fled the plantation, they were specifically prohibited by royal decree. Their political potential as a weapon of the slaves and a nefarious threat to the planters was made explicit upon the arrest of the Maroon leader Medor. "If the blacks commit poisonings," he told his captors, "the end purpose is to gain freedom."

By the last years of French rule, it was patently clear to all that the greed of the entire system had set the colony on a path of self-destruction. Only the potential for massive profits could possibly have numbed the whites to the imminent disaster. As absentee planters scrambled to increase their holdings, the borders of the plantations touched, and then to meet the rising demand for coffee, in particular, rose higher and higher into the mountains, displacing bands of Maroons and ironically forcing more and more of them to depend on pillage. The voracious consumption of labor, meanwhile, doubled the population of slaves in a mere fifteen years. How long could the whites possibly have expected to control close to a half-million blacks, the vast majority of them born in Africa and steeped in a military tradition of their own that had spread kingdoms across half a continent?

Like Macandal, the slaves plotted their final revolt with care. As their network spread and desertions swelled the ranks of the Maroons, the night air reverberated with the sounds of mutiny. In 1786 an informant reported clandestine meetings of two hundred or more slaves being held on the plantations. A contemporary document states "a great deal has been said of slave superstitions and of their secret organizations, and the scheming and crimes for which they provided pretext—poisonings, infanticide— . . . whites were not admitted to these secret meetings, and legal documentation was usually held secret or destroyed."

From these nocturnal assemblies and within the passion of the vodoun rituals, the idea of liberty spread. Maroon bands grew in number as well as size, and the names of their leaders—Hyacinthe, Macaya, Romaine La Prophétesse—spread through the ranks of the slaves. The flash point came in the summer of 1791, and the spark was cast at a vodoun ceremony attended by delegates from every plantation on the northern plain.

The historic gathering was invoked by the Maroon leader Boukman Dutty, and held on a secluded knoll at Bois Caiman near Morne-Rouge. There on August 14, 1791, beneath the spindly branches of a frail acacia, with the wind twisting the ground and the jagged lightning crashing on all sides, an old woman stood transfixed by the night, quivering in the spasms of possession. The voices of Ogoun the Warrior, the god of fire and the metallurgical elements, called for the cutlass, and with a single blow severed the head and spilt the foamy blood of the black pig of Africa. The leaders were named—Boukman himself, Jean François, Biassou, and Celestin—and one by one the hundreds of slaves swore allegiance. Boukman stood up, and in a voice that matched the fury of the wind cried out, "God who made the sun that shines on us from above, who makes the sea to rage and the thunder roll, this same great God from his hiding place in the clouds, hear me, all of you, is looking down upon us. He sees what the whites are doing. The God of the whites asks for crime; ours desire only blessings. But this God who is so good directs you to vengeance! He will direct our arms, he will help us. Cast aside the image of the God of the whites who thirsts for our tears and pay heed to the voice of liberty speaking to our hearts." Thus was sealed the pact of the final revolt, in the shadow of the loa and with the blood of the sacrifice.

Two days later the first plantation burned, and then on the night of August 21 the slaves at five plantations around Macandal's old territory rose up and moved on the center of Limbé. By morning Acul was in flames, Limbé destroyed, and throughout the next day the uprising rode the sound of the conch trumpets as one by one each settlement in the north fell—Plaine du Nord, Dondon, Marmelade, Plaisance. In a single night of indescribable horror, a thousand whites were strangled and two thousand sugar and coffee mills destroyed. For days dark columns of smoke rose beyond a wall of flames that isolated the entire northern half of the colony. Fire rained from clouds of burning straw torn from fields and swept up by the fireballs. Ash coated the sea, and the image of an entire land aflame reddened clouds as far away as the Bahamas.

After the first weeks of the uprising, as the frenzy of destruction gave way to skirmishes and then out-and-out battles with the colonial militia, the ranks of the slaves coalesced around certain leaders, who in turn drew their inspiration from the gods. In the western province Romaine La Prophétesse marched to the music of drums and conch shells, behind an entourage of houngan chanting that the weapons of the whites, their cannon and muskets, were bamboo, their gunpowder but dust; his personal guard carried only long cowtails blessed by the spirits and thus capable of deflecting the bullets of the whites. Sorcerers and magicians composed the staff of Biassou, and much of his own tent was devoted to amulets and sacred objects and devotions. In his camps great fires flared by night as naked women invoked the spirits, singing words known only in "the deserts of Africa." Biassou walked in triumph, exalting his people, telling them that if they were fortunate enough to fall in battle, they would rise again from the hearth of Africa to seed their ancient tribes. A contemporary report from Cap Francis suggests that the black women of the capital went out at night, singing words unintelligible to the whites. For some time they "had adopted an almost uniform dress, around their bellies wearing kerchiefs in which the color red dominated. . . . The Voodoo King had just declared war (they said) and accompanied by his Queen dressed in red scarf and agitating the little bells decorating the box containing the snake, they marched to the assault of the colony's cities."

If vodoun charged the revolt, the tactics and organization of the rebel bands came directly from the precedent established by the Maroons. With increasing boldness the rebels poured out of the hills ravaging plantations, disrupting communication, and plundering supply trains. Raiding by night, they left in their wake fire, poison, and corpses, before retreating at daybreak into every inaccessible gorge and ravine in the land. There they lived as they had always, protected by palisades and rings of sentinels, and most critically by the potent magic of their sorcerers. A French force of some twenty-four hundred troops dispatched in early February 1792 to invade and destroy rebel camps reported being "astonished to see stuck in the ground along the route large perches on which a variety of dead birds had been affixed. . . . On the road at intervals there were cut up birds surrounded by stones, and also a dozen broken eggs surrounded by large circles. What was our surprise to see black males leaping about and more than 200 women dancing and singing in all security. . . . The Voodoo priestess had not fled . . . she spoke no creole. . . . Both the men and the women said that there could be no human power over her . . . she was of the Voodoo cult." The leader of the French expedition also encountered the cloak of secrecy that continued to protect the rebels as it had the Maroons. From a woman initiate he learned that "there was a password but she would never give it to me. . . . She gave me the hand recognition sign: it was somewhat similar to that of the Masons. She told me this as a secret, assuring me . . . I would be killed or poisoned if I tried to penetrate the great mystery of the sect."

Perhaps unfortunately for the ex-slaves, they were not alone in their quest for freedom. For some time tension had been growing between the white planters and the enfranchised mulattos. The latter, though equal in numbers to the whites, and entitled by law to all the privileges of French citizens, were, in fact, treated as a class apart. Social discrimination of the crudest sort had made them resentful, and at the same time their energetic exercise of their right to property had made them exceedingly wealthy and powerful; this was a dangerous combination for the French, especially once fanned by the rebellious ideas that had grown out of the American and French Revolutions. Even before the uprising of the slaves, a force of mulattos had marched on the government demanding full implementation of the accords worked out by the revolutionary assemblies in Paris. Their incipient revolt failed, and the leaders were brutally tortured, but it was the first incident in a struggle that would, in time, pit whites against mulattos, whites against mulattos allied with blacks, and mulattos against whites working sometimes with and sometimes against the interests of the black ex-slaves. The result was a reign of confusion, chaos, and destruction that after two years left the French in control of the cities, the blacks firmly entrenched in the countryside, and the mulattos still caught in between.

In February 1793 the Haitian revolution was profoundly affected by the war that broke out in Europe between Republican France and England allied to Spain. To the problems of the colonial administration, already preoccupied by political events in Paris, stunned by the uprising of the slaves, and torn asunder by the power struggle between mulatto and white planters, was now added the threat of an enemy army moving overland from the Spanish colony of Santo Domingo. The Maroons predictably aligned themselves with the invaders, and their leaders Biassou and Jean François—Boukman by then being dead—became officers in the Spanish army, as did Toussaint, an ex-slave who had not left bondage until some months after the uprising, but who was about to begin his meteoric rise to power.

The French, hard-pressed even before the Spanish invasion, were now forced to negotiate with the Maroons, and in the summer of 1793, under pressure from the rebel leaders, the colonial administration officially abolished slavery. But the Maroons soon recognized the proclamation for what it was—no more than a desperate attempt to defuse the explosive potential of the blacks, while retaining intact the essential structure of the colonial economic order. Biassou and Jean François rejected the French offer and defiantly remained allied to the Spanish, who had promised unconditional emancipation. But for reasons that would only come clear in the light of subsequent events, Toussaint shifted his allegiance back to the French. Then, at a time when he remained unquestionably less powerful than the prominent Maroon leaders, Toussaint and his forces ambushed the camps of Biassou and Jean François, killing many of their followers and turning those that survived over to the French authorities.

Toussaint moved quickly to consolidate his position, and by 1797, having defeated two foreign armies as well as a rival mulatto force, this ex-slave, by now christened Toussaint L'Ouverture, emerged as the absolute ruler of all Saint Domingue. He then set to work to restore under French rule the order and prosperity of the colony. And of what type of colony, there could be no doubt. The people, though nominally free, were in fact compelled to work in a manner reminiscent of exactly the system they had fought so desperately to overthrow. Toussaint's autocratic decrees prohibited movement between plantations and ordered those without urban trades to the fields, where they labored under military supervision. True, he eliminated the worst excesses of the colonial brutality, but the essential structure of the plantation system remained just as the French had intended. As for the traditional beliefs of the people, Toussaint as a devout Roman Catholic had no interest in what he considered the pagan beliefs of Africa.

If there were any doubts as to the ambitions of the new black military elite, they were to be quelled by events that took place following the invasion of Leclerc's French army in 1801, the betrayal of Toussaint by Napoleon, and his ignoble deportation to France. The French, of course, had never intended to tolerate black rule in the colony, and even as Toussaint struggled to rebuild the plantations, plans for his overthrow were laid in Paris. Napoleon, himself, clearly identified his real military enemy in the colony and ordered his brother-in-law to concentrate his efforts, following the deportation of Toussaint, on destroying the remnants of the Maroon bands. As was predicted in Paris, it was a task that the French commander was able to assign to local generals that had so recently served under Toussaint—in particular Dessalines and Christophe, who by historical record applied themselves and their troops willingly to the slaughter.

Still the Maroons resisted, and as the pressure mounted their forces were in place to receive both black freedmen and mulattos into the final alliance that would wage the war of independence. There were a number of pitched battles, but there is little doubt that the tried-and-proven tactics of the Maroons—raids, fire, poison, ambush—provided the margin of victory. Yet in the final defeat of the French, the Maroons were as soon forgotten as they had been after the wars of Toussaint L'Ouverture. The dictatorial rulers that emerged in the vacuum of independence lost no time in revealing the depths of their dedication to equality or liberty. Christophe—a former slave who had fought by the side of Toussaint, taken part in the deceitful raid on the forces of Biassou and Jean François, and then temporarily joined hands with the French armies of Leclerc—became the ruler of the northern half of the country in 1806. He declared himself king, and in the ways of kings he wasted the lives of twenty thousand subjects to build himself an opulent palace and a fortress that would never fire a shot.

With Christophe the betrayal of the Maroons appeared to be complete, but their struggle, in fact, continued. Those of their leaders who were still alive, and who had been instrumental in maintaining the revolt, had long grown accustomed to deceit, and they were, if nothing else, resilient. In the early years of the republic, as they faced new oppressors, they also found a new and remarkable means of protecting their freedom.

Some fifty years ago, Zora Neale Hurston, a young American black woman and a former student of the great ethnographer Franz Boas, stumbled upon an extraordinary mystery while preparing for her first field trip to Haiti. In what was at the time perhaps the only reliable monograph on the vodoun society, she read that in the valley of Mirebalais there were secret societies that terrorized the local inhabitants. According to the author, the noted Africanist Melville Herskovits, these clandestine organizations were called together at night by the beating of one rock against another in a manner reminiscent of the Zangbeto, a secret society he had studied in Dahomey. So feared were these Haitian groups that only with great difficulty had Herskovits obtained the names of two. One was the *bissago*, whose members appeared at night "wearing horns and holding candles," and the other was *Les Cochons sans Poils*—the Pigs without Hair. Both societies were believed capable of changing their members into animals and sending them into the night to spread evil.

Although this was the first report of secret societies in Haiti that Hurston had come across, the persistence of such a prominent West African cultural trait had not surprised her. For Zora Neale Hurston, born at the turn of the century in a small all-black village in Florida, had come at an early age to sense the African roots of her people. Her father was a Baptist preacher, and behind the frenzy of his service, the wailing gospel hymns, the sermons and spirit possession, she had perceived even as a young girl what she would recognize as the raw power of African worship. Nine years old when her mother died, Zora Neale promptly left home and drifted north, working as a maid for a traveling theatrical troupe before eventually ending up in Baltimore, where she finished her schooling. Luck rode with her, and her interest in literature and folklore led her to Howard University, and hence to a scholarship at Columbia. There she met Boas, who became her "Papa Franz," her confidant, intellectual mentor, and strongest supporter.

At the time Franz Boas was in the process of revolutionizing the field of anthropology. In an era when British social anthropology was still an explicit tool of imperialism, he rejected arbitrary notions of progress and evolutionary theories that invariably placed Western society at the top of a social ladder. Instead he championed the need to study cultures because of their inherent value. Every culture possessed a certain logic, he would say, and they appeared peculiar to the outsider only because he or she did not understand that logic. More than anyone before him, Boas saw anthropology as a calling, an opportunity to herald the wonder of cultural diversity, while revealing the intricate weave of the human fabric that binds us all together. In the spirit of her teacher, Zora Neale Hurston had been one of the first to undertake scholarly research in the field of Afro-American folklore. At this time when racism was fashionable, she had written boldly that the "Hoodoo doctors [of the American South] practice a religion every bit as strict and formal as that of the Catholic Church."

Inspired by Boas's stress on the need for fieldwork, Zora Neale Hurston developed her own style, to say the least. Packing a pearl-handled pistol, she roamed the dusty backroads of the deep South in a beat-up Chevrolet, seeking out the hoodoo doctors, guitar players, and storytellers, herself playing just about as many roles as there were characters in the stories she collected. Sometimes they thought she was a bootlegger's woman on the run, or a widow looking for a man. And when she sped away singing some bawdy song, they were certain she was a vaudeville star looking for new material. This amazing woman decked out in her beret and cheap cotton dresses, with all she possessed stuffed into a flimsy suitcase, reached into every bayou and woodlot in the South, and when she got there she carried to its limits what anthropologists stiffly call participant observation. Once to satisfy a hoodoo priest she had to steal a black cat, then kill it by dropping it into boiling water; after the flesh fell away, she was instructed to pass each bone through her teeth until one tasted bitter. During an initiation ceremony in New Orleans, she had to lie naked for sixty-nine hours on a couch with a snakeskin touching her navel; at the tolling of the seventieth hour, five men lifted her from the floor and began a long ritual that culminated with the painting of a symbolic streak of lightning across her back. Then the hoodoo priest passed around a vessel of wine mixed with the blood of all present. Only by sharing in this ritual drink did Zora Neale become accepted by the cult.

It was this spirit of adventure combined with a passionate desire to continue her investigations and promote vodoun as a legitimate and complex religion that drew Hurston to Haiti. There is little doubt that by the time she read of the secret societies her proposed journey had become something of a personal crusade. For some time American and European foreign correspondents had indulged their readers' perverse infatuation with what was known as the Black Republic, serving it up garnished with every conceivable figment of their imaginations. For Americans, in particular, Haiti was like having a little bit of Africa next door, something dark and foreboding, sensual and terribly naughty. Popular books of the day, with such charming titles as *Cannibal Cousins* and *Black Bagdad,* cast the entire nation as a caricature, an impoverished land of throbbing drums ruled by pretentious buffoons and populated by swamp doctors, licentious women, and children bred for the cauldron. Most of these travelogues would have been soon forgotten had it not been for the peculiar and by no means accidental timing of their publication. Until the first of this genre appeared in 1880—Spenser St. John's *The Black Republic,* with its infamous account of a cannibalistic "Congo Bean Stew"—most books that dealt with vodoun had simply emphasized its role in the slave uprising. But these new and sensational books, packed with references to cult objects such as voodoo dolls that didn't even exist, served a specific political purpose. It was no coincidence that many of them appeared during the years of the American occupation (1915–1934), and that every Marine above the rank of captain seemed to manage to land a publishing contract.

There were many of these books, and each one conveyed an important message to the American public—any country where such abominations took place could find its salvation only through military occupation. Zora Neale Hurston, as much as anyone, was aware of these slanderous accounts, and she was quick to realize how the material she had collected in the American South could be exploited if placed in the wrong hands. Thus it was not only a trained but a judicious eye that now was turned on the Haitian secret societies.

Even within her first weeks in Haiti, as she moved at will through the streets of Port-au-Prince, Zora Neale heard the whispers running along the front edge of the night. They came at first alone, single incidents without any obvious pattern. First there was the drum that sounded one Thursday above her village, a rapid staccato beat, highly repetitious and unlike any rhythm she had heard in a hounfour. She woke up her housegirl and suggested they investigate, much as they had done many times before. But instead of the eager companion she had grown accustomed to, Zora Neale beheld for the first time a trembling child, unwilling to step beyond the threshold of their doorway. Then some weeks later, again at night, Hurston was disturbed by the acrid smell of burning rubber in her yard. When she questioned the man responsible, he apologized profusely but explained that the fire was necessary to drive away those who planned to take his child, the *Cochons Gris*, or Gray Pigs, who he claimed ate people. Apparently he had already seen them parading by the house, figures draped in red gowns and hoods. Yet a third incident occurred on a sailboat en route to the island of La Gonave, when she encountered a force of militia moving in to suppress some society based in a remote region of the island. No details were provided, just a few words followed by a hush of uneasiness, which Hurston interpreted as fear. Finally back in the capital, a contact led her to the house of a houngan in the Bel Air slum, and there she saw a temple unlike anything she had known. At the center of the sanctum was an immense black stone attached to a heavy chain, which in turn was held by an iron bar whose two ends were buried in the masonry of the wall. As she stood before the stone, her host handed her a paper, yellow with age and patterned with cabalistic symbols, and a *"mot de passage"*—password of the secret society known as the Cochons Gris.

Following these and other leads, Zora Neale over the course of several months managed to put together an astonishing portrait of the Haitian secret societies. According to her informants they met clandestinely at night, called together by a special high-pitched drumbeat; members recognized each other by means of ritualized greetings learned at initiation, and by identification papers—the passports. In a vivid description of a nocturnal gathering, she mentions a presiding emperor accompanied by a president, ministers, queens, officers, and servants, all involved in a frenzied dance and song ritual that she likens to the "sound and movement of hell boiling over." The entire throng formed a procession and marched through the countryside, saluting the spirits at the crossroads, picking up additional members before convening in the cemetery to invoke Baron Samedi, the guardian of the graveyard, and beg him to provide a "goat without horns" for the sacrifice. With the spirits' permission, scouts armed with cords made from the dried entrails of victims combed the land for some traveler foolish enough to be still journeying after dark. If the unfortunate wayfarer was unable to produce a passport indicating his membership in a society, and thus his right to move by night, he was taken to be punished.

Unfortunately, it appears from her writings at least that Hurston was never able to attend one of these gatherings herself. Perhaps because of this she was unable to get beyond the public face of the societies, an image that cast them in the words of a later ethnographer as "bands of sorcerers, criminals of a special kind." A member of the mulatto elite, for example, told Hurston:

We have a society that is detestable to all the people of Haiti. It is known as the Sect Rouge, Vinbrindingue and Cochons gris and all these names mean one and the same thing. They are banded together to eat human flesh. . . . These terrible people were kept under control during the French period by the very strictures of slavery. But in the disturbances of the Haitian period, they began their secret meetings and were well organized before they came to public notice. . . . It is not difficult to understand why Haiti has not even yet thoroughly rid herself of these detestable creatures. It is because of their great secrecy of movement on the one hand and the fear that they inspire on the other. . . . The cemeteries are the places where they display the most horrible aspects of their inclinations. Some one dies after a short illness, or a sudden indisposition. The night of the burial the Vinbrindingues go to the cemetery, the chain around the tomb is broken and the grave profaned . . . and the body spirited away.

Yet who were these secret societies? Zora Neale Hurston never really says. She concludes simply by emphasizing that they are secret, and that the very lives of their members depend upon the confidence of the group. But she also, rather remarkably, describes what she could not have realized was the primary method of zombification. "There is," she writes, "swift punishment for the adept who talks. When suspicion of being garrulous falls upon a member, he or she is thoroughly investigated, but with the utmost secrecy without the suspect knowing that he is suspect. But he is followed and watched until he is either accounted innocent or found guilty. If he is found guilty, the executioners are sent to wait upon him. By hook or crook, he is gotten into a boat and carried out beyond the aid and interference from the shore. After being told the why of the thing, his hands are seized by one man and held behind him, while another grips his head under his arm. A violent blow with a rock behind the ear stuns him and at the same time serves to abrase the skin. A deadly and quick-acting poison is then rubbed into the wound. There is no antidote for the poison and the victim knows it."

It wasn't until almost forty years later that a young Haitian anthropologist by the name of Michel Laguerre managed to begin to answer some of the questions raised by Hurston's fieldwork. In the summer of 1976, Laguerre met a number of peasants who had been invited to join the secret societies, but who later had converted to Protestantism and hence were willing to talk. There were, according to these informants, secret societies in all parts of the country, and each one maintained control of a specified territory. Names varied from region to region but included Zobop, Bizango, Vlinbindingue, San Poel, Mandingue, and, most interestingly, Macandal. Membership was by invitation and initiation, open to men and women, and was strictly hierarchical. Laguerre verified the existence of passports, ritual handshakes and secret passwords, banners, flags, and brilliant red-and-black uniforms, as well as a specialized body of spirits, songs, dances, and drumbeats. He also noted the central importance of the cyclical rituals performed to strengthen the solidarity of the group: gatherings that occur only at night, that begin with an invocation to the spirits and end with the members in procession, flowing beneath the symbol of the society, the sacred coffin known as the *sekey madoulè*.

But according to Laguerre the function of the secret societies was unlike anything reported by Hurston. In no way could they be considered criminal organizations. On the contrary, he described them as the very conscience of the peasantry, a quasi-political arm of the vodoun society charged above all with the protection of the community. Like the secret societies in West Africa, those of Haiti seemed to Laguerre to be the single most important arbiter of culture. Each one was loosely attached to a hounfour whose houngan was a sort of "public relations man" acting as a liaison between the clandestine society and the world at large. In fact, so ubiquitous were the societies that Laguerre described them as nodes in a vast network that, if and when linked together, would represent a powerful underground government capable of competing head-on with the central regime in Port-au-Prince. And of the origin of these secret societies, Michel Laguerre had no doubts.

It was his view that in the aftermath of the war of independence, and in the face of a history of betrayal at the hands of the military leaders, the struggle of the Maroons had continued. In certain parts of the country Maroon bands had persisted as late as 1860 but on the whole, as the ex-slaves took to the land and the vodoun society was born, the role of the Maroons was transformed from fighting the French to resisting a new threat to the people—an emerging urban economic and political elite distinguished not by the color of their skin but by the plans they harbored for both the land and labor of the peasants. And whereas the war with the French could be won, the new struggle promised to be a permanent feature of peasant life. From overt and independent military forces, the Maroons went underground and became a clandestine institution charged with the political protection of the vodoun society. Thus were conceived the immediate predecessors of today's secret societies. As the Maroons fought for liberty, the contemporary secret societies, according to Laguerre, "stand strong to keep safe the boundaries of power of their local communities and to keep other groups of people from being a threat to their communities."

If Laguerre was correct—and his was the only explanation that made sense to me—the implications were clear. From the leading medical authorities in Port-au-Prince I had been told that zombification was a criminal activity that had to be exposed and eradicated for the collective good of the nation, but now a quite different scenario presented itself. In the minds of the urban elite, zombification might well be criminal, but every indication suggested that in the vodoun society it was actually the opposite, a social sanction imposed by recognized corporate groups whose responsibility included the policing of that society. Zombification had always struck me as the most horrible of fates. Now, if what I was beginning to think was true, I realized that it had to be. After all, what form of capital punishment is pleasant?

I knew from my own research that in at least some instances the zombi powder was controlled by the secret societies, and a knowledge of poisons and their complex pharmacological properties could be traced in a direct lineage from the contemporary societies to the Maroon bands, and beyond to the secret societies of Africa. There was no doubt that poisons were used in West Africa by judicial bodies to punish those who broke the codes of the society, and Hurston had suggested that the same sort of sanction was applied among the secret societies of Haiti. I had every reason to believe that Ti Femme and Clairvius Narcisse had received a poison, and that at the time of their demise they were both pariahs within their respective communities. Narcisse's transgression was directly related to access to land, precisely the sort of issue that, according to Laguerre, would attract the attention of the secret society. By his own account, Narcisse had been taken before a tribunal, judged, and condemned. The possible link to the tribunals of West Africa was obvious. What's more, Narcisse had referred to those who had judged him as the "masters of the land," and had implied that he would only encounter more trouble if he created further problems himself. Finally, Max Beauvoir had told me directly that the answer to the zombi mystery would be found in the councils of the secret societies.

If my identification of the zombi poison revealed a material basis for the phenomenon of zombification, the work of Michel Laguerre combined with the information provided by Max Beauvoir and others finally suggested a sociological matrix. And once again, the indomitable Zora Neale Hurston provided a critical clue.

In October 1936, a naked woman was found wandering by the roadside in the Artibonite Valley. She directed the authorities to whom she was taken to return her to her family land, and there she was identified by her brother. The woman's name was Felicia Felix-Mentor. Like Ti Femme she was a native of Ennery, and twenty-nine years before she had suddenly taken ill, died, and been buried. Death certificates and the testimony of her ex-husband and other family members seemed to support her account. When found she was in such a wretched condition that the authorities, after her identification in Ennery, placed her in a hospital in Gonaives, and it was there that Zora Neale Hurston found her. In Hurston's own words, the woman was a dreadful sight— a "blank face with dead eyes" and eyelids "white as if they had been burned with acid."

Hurston remained with the reputed zombi for only a day, but she came away from the encounter quite able to tell the world just about everything there was to know about the phenomenon. Unfortunately, nobody believed her. For example, she later wrote that she and the attendant physician "discussed at great length the theories of how zombies came to be. It was concluded that it is not a case of awakening the dead, but a matter of a semblance of death induced by some drug known to a few. Some secret probably brought from Africa and handed down generation to generation. The men know the effects of the drug and the antidote. It is evident that it destroys that part of the brain which governs speech and will power. The victim can move and act but cannot formulate thought. The two doctors expressed their desire to gain this secret, but they realize the impossibility of doing so. *These secret societies are secret* [italics mine]."

When Hurston published this hypothesis in 1938 in *Tell My Horse*, it was ignored in the United States, while in Haiti it earned her the scorn of the intellectual community. It wasn't that the notion of a poison seemed impossible. On the contrary, the belief in the poison was so common that virtually every subsequent student of Haitian culture would make some reference to it. Alfred Métraux's suggestion that the "hungan [sic] know the secret of certain drugs which induce a lethargic state indistinguishable from death" was typical. And though most anthropologists remained equivocal, the Haitians themselves recognized the existence of the poison with such assurance that it was specifically referred to in the Penal Code.

Hurston's problem was less one of credibility than of timing. Her report appeared just in the period when Haitian social scientists trained in the modern tradition were most anxious to promote the legitimacy of peasant institutions. These intellectuals were still smarting from the sensational publications that had emanated from the United States, which in their minds had both slanderously misrepresented the Haitian people and rationalized the American occupation. The subject of zombis, which had figured so prominently in these books, as it would later in low-budget Hollywood movies, was simply anathema to them. Unworthy of serious consideration, the embarrassing phenomenon was dogmatically consigned to the realm of folklore. Zora Neale Hurston, meanwhile, whose insight, if encouraged and pursued, might have solved the zombi mystery fifty years ago, bore the brunt of her colleagues' contempt.

In defense of her critics, it must be said that her case for the poison hypothesis was suspect on several grounds. First, many informants insisted that the actual raising of a zombi depended solely on the magical power of the bokor. Secondly, despite the provocative case of Felicia Felix-Mentor, no physician had examined an indisputably legitimate case. And finally, of course, no one had penetrated the cults to obtain a sample of the reputed poison. Hurston did not help her position by concluding that "the knowledge of the plants and the formulae are secret. They are usually kept in certain families and nothing will induce the guardians of these ancient mysteries to divulge them." Here, unfortunately, Hurston was unduly influenced by the warnings she had received from Haitian medical authorities, who had obviously sought the formula in a completely inappropriate manner. In one instance, for example, a zealous physician had used his connections to have a bokor imprisoned on no charges and threatened with a long term unless he divulged the secret. Not surprisingly, the recalcitrant old man refused, saying it was a mystery brought over from Guinée, which he would never reveal. From these same authorities Hurston received the dire warnings that may have prevented her from pursuing the mystery. "Many Haitian intellectuals," she was told, "are curious, but they know that if they dabble in such matters, they may disappear permanently." If she persisted in her desire to contact the secret societies directly, she was told, "I would find myself involved in something so terrible, something from which I could not extricate myself alive, and that I would curse the day that I had entered upon my search."

Zora Neale Hurston was a woman of uncommon courage, but she worked as a pioneer in a complete vacuum that left her no option but to heed the words of her Haitian colleagues. But if Michel Laguerre was correct and the secret societies represented a legitimate political and judicial force in the vodoun society, it had to be possible to contact them safely. Only by doing precisely that could the final aspect of the zombi mystery be solved. And in Herard Simon, I had someone who could take me there.

12

Dancing
in the Lion's Jaw

MY REFLECTIONS from the vantage of New York and Cambridge trembled within me like a lodestar, pulling me back to Haiti. The desire to understand the connection between the creation of zombis and the secret societies had, by the late fall of 1982, extended into an ambition to penetrate the groups themselves. Only through this final step could I get beyond the public face, could I understand why zombis were made.

It was a potent idea, and one that, if I were to accept the counsel of my advisors, was fraught with danger. After all, I was in effect asking who actually ruled rural Haiti. Nathan Kline, though remaining wed to the notion of raising a zombi from the grave, knew enough about Haiti to appreciate immediately the ramifications of my proposed venture, and in a series of meetings held throughout the early winter offered his total endorsement and support. Heinz Lehman remained cautious, and both he and the producer David Merrick, who I had since discovered was the principal financial backer of the project, echoed the concern of various anthropologists who had suggested that the most distinct and likely outcome of such an adventure was that I would get myself killed. I reassured them that based on the information I had, the dangers had been exaggerated. Finally it was agreed that I would be given full financial support, with the one stipulation that I give priority to arranging for the medical documentation of the resurrection of a zombi. But then, soon after the last of our meetings, circumstances beyond our control eclipsed any plans for my return. In mid-February 1983, Dr. Nathan Kline died unexpectedly while undergoing routine surgery at a New York hospital. Within forty-eight hours David Merrick suffered a debilitating stroke that left him incapacitated and severed his interest in the project. Thus, as the year turned through a winter of tragedy and rapid change, the chance to study the secret societies slipped temporarily from my grasp.

It was a full year later before Haiti would again claim me, this time without backers, and with many things changed. Not the country, of course. Driving once more through the streets of Port-au-Prince, past the gingerbread houses with their tall palms thrust to the sky, past the stagnant pools of sewage by the side of the Truman Boulevard and the men from the public works, a mantle of green slime clinging to their thighs, I remained fixed as ever by the words of the stranger at the Hotel Ollofson: "Haiti will remain Haiti as long as the human spirit ferments." Still, along with the easy happiness I had come to associate with the country, I was aware of a new and perhaps less superficial sensation—that sense of familiarity and alienation that comes to one who knows a place well, but who can never hope to become a part of it.

Among my old contacts I would soon discover that time allowed us to expose new facets of ourselves, revealing contradictions previously unseen. On the surface, however, very little seemed to have changed. In Saint Marc, Marcel Pierre had reached giddy heights of fame as a result of the BBC documentary having been aired on national television, and he had recently been seen wandering the halls of the local hospital parroting the words of the reporter and insisting that he was not serving evil, but the future medical needs of all humanity. His exhilaration would soon be checked by the brutal illness of his favorite wife, who was slowly bleeding to death from a malignant tumor in her uterus. Max Beauvoir was still his ebullient, debonair self, only far poorer due to the virtual cessation of the tourist trade in the wake of the AIDS scare. As for myself, the death of my benefactor had brought a marked change in my financial status, and what little money I had soon went into buying blood for Marcel's wife. The Haitians responded to all of this in a most unexpected way. When I had been flush with Kline's money, they had done what they could to relieve me of the burden; now that I was limited to my own comparatively scant resources, they asked of me almost nothing.

But of all of us who had joined hands in this unlikely drama, it was Rachel who had most truly changed. In the fall of 1982 she had begun her studies in anthropology at Tufts University, and her exposure to the United States had reaffirmed her identity among her own people. In a very real way, she had discovered her sense of place, and the free, undecided spirit I had known in Haiti had become committed. When I told her of my plans, she wanted to pursue the mystery with me. She contacted her academic advisor and arranged to receive credit for the time.

We knew we would have to start with Herard Simon.

In the late afternoon the air had been stifling, but then the rain began as it so often does in the early summer evenings, first in wild gusts and then more steadily in broad sheets that filled the horizon. As suddenly as it began the rain had stopped, leaving something ominous in the steely night wind that blew through the streets of Gonaives. A partial blackout had cut off the electricity, and kerosene lamps flickered in many quarters, casting a wan light. The waterfront, though, had been spared; a large crowd huddled beneath the marquee of the movie house where Herard Simon was supposed to be waiting. Herard loved the movies, and this dilapidated theater was one of his favorite conduits to the outside world. He took his films in pieces, dropping in on a whim, almost never watching from start to finish—a peculiar habit but one perfectly suited to Gonaives's only theater, where American films are dubbed into garbled French that crackles in the ears of people who understand only Creole.

A legless man, smiling and propped on a trolley, greeted us with a message. I followed Rachel around the corner by the theater and up a side street to the home of the ex–police chief, where we paused long enough to pick up yet another message that led us first back to the waterfront and finally to the outskirts of the city to the Clermezine nightclub. There Herard's wife Hélène received us warmly, and while we waited for her husband she treated us to an exhilarating account of the day's activities in the market. As we sat in the darkness with the cloying richness of her perfume mixing with the damp yet wholesome air, she spun a story of immense bathos, an agony of linked phrases repeated over and over so that the thrill of her experience might last forever, even though nothing had really happened and by tomorrow all would be forgotten. It was a typical, however extraordinary, performance, one that might have gone on for hours had it not been interrupted by the arrival of Herard. Though it had been well over a year since we last saw each other, we met as friends do after a passage of days. Herard deflected my enthusiastic greeting, and after I had quickly exhausted a few trivial bits of news, I realized that it was silence more than words that would define our relationship. In the West we cling to the past like limpets. In Haiti the present is the axis of all life. As in Africa, past and future are but distant measures of the present, and memories are as meaningless as promises.

Still, I had come back, and I sensed that this meant something to him. But in the murky light his moonlike face remained more inscrutable than ever, his presence conveying the self-assurance of a man fully capable of mixing stars with sand, of carrying lightning in a pocket. From far to the west came the rumbling of thunder, and closer the sound of wind stiffly clicking the leaves of the almond trees that grew over the compound. And then his familiar rasping laughter, signaling his approval of the gifts I had brought. The last time I had seen him I had told him of an interim assignment I was planning into the Amazon, and I had asked him what he wanted from the other side of the water. "Something mystical," was his reply. It was a tall order coming from him, but one that I had tried to fill by bringing him an ocelot pelt and some vertebrae of a large boa constrictor.

"Did you eat the meat?" he asked as Rachel and I slowly unfolded the pelt.

"They say it is forbidden," I answered honestly.

"The whites say this?"

"No, the Indians."

"Good. You see," he said, turning to Rachel, "it is as I told your father. Someone of such wildness learns nothing from his elders. So he goes to the wild places to be among the leaves. Now he appears again among the living because leaves are not enough."

I followed his logic only enough to see that Herard had found a comfortable image for me, untrue but meaningful for him. Since I fitted none of his categories and defied his common image of foreigners, he had forged a new category, a composite not unlike a collage he might have made up of scattered impressions cut from a dozen grade B movies. The jungles, the Amazonian myths and tribes I had spoken of, the animals I had described, some photographs I had shown him: in the end I was something wild, not a white, and that was all there was to it.

"But that's not why he is here," Rachel said. "He has come back because we have come together, because . . ."

Herard lifted his hands before his face, then rose painfully to his feet muttering an undercurrent of groans and unintelligible words.

"Your father has told me," he said finally. "Rachel, do you think this is a game? Bizango is diabolic. It is not what you think."

"But there are those who say the Bizango rite is life itself."

Herard had not expected such a prompt and audacious reply from the Rachel he had known, and for a quick moment a look of baffled vexation came over him. "They can say what they want. The ritual speaks the truth. Listen to the songs. What do they say? None of them says 'give me life.' And when they put the money in the coffin, what do they sing? 'This money is for the *djab*,' the devil, or 'Woman, you have two children, if I take one you'd better not yell or I'll eat you up!' The songs have only one message—Kill! Kill! Kill!" Rachel began to say something, but Herard was not to be interrupted. "To do a good service in the Bizango you must do it in a human skull. And it can't be a skull from beneath the earth, it must be a skull they prepare. Their chalice is a human skull. What does this tell you, child?"

"It is something that must be done." Rachel was unrelenting, her voice untouched by fear.

"Then let me tell you what will happen. When an outsider intrudes on the society, when he tries to enter the Bizango, he receives a coup l'aire or a coup poudre. Do you want to see beasts fly? Yes, I suppose you do. Well, if you are lucky they shall only frighten you and tie you to the poteau mitan while twenty Shanpwel with knives dance around you. On the side they'll have a pot of oil, with cooking meat floating on the surface. Only they'll have a finger in it, and you will not know if the meat comes from your mother. Then they judge you, and you pray that the president says you're innocent."

"But we shall be."

He grabbed my hand, holding it close to his face until I could feel his breath. "Not him!" he snorted.

"No Haitian reads the color of a man's skin."

"Girl, stop this foolishness. The Americans stole the country once. In the days of the Father they tried to take it again. No foreigner walks under the cover of the night."

"Unless you take us."

"Never. Rachel, your days are young, and Wade must still serve the loa. Bizango is djab, it is evil, and you must not begin it."

"We only want to see what they do."

Herard said nothing. He was a man long unaccustomed to argument. Usually when others had finished talking, he would declare his will in a few flat phrases and wait calmly for obedience. But tonight, oddly, Rachel had the last word.

Our disappointment in Herard Simon's rebuff aside, we soon found, to my surprise if not Rachel's, Haitians who were more than willing to speak about the Bizango or Shanpwel, terms that many used interchangeably. Within a matter of days of our visit to Herard, Rachel and I had heard people accuse the secret societies of just about every conceivable amoral activity from eating children to transforming innocent victims into pigs. From everything we could gather the public face of the Bizango was still as nefarious as anything that had been reported in the popular or academic literature. It was therefore with special interest that we listened to the account told us by a young man from the coastal settlement of Archaie, a fellow named Isnard who was twenty-five when he entered the Bizango in 1980.

Since his youth Isnard had been warned against going out at night, but one evening when he again heard the drums of the society, and while his mother thought he was asleep, he slipped out of the lakou and followed the sound to a not-distant compound. At the gate he was met by a sentinel, who turned out to be a friend, and while they were speaking a man identified as the president of the local society came out to share a drink with the sentinel. The president, also a friend, invited Isnard to enter. That night two societies were meeting, and the bourreau, or executioner, of the visiting society was an enemy of Isnard's. Immediately he gave the order for Isnard to be "caught." A call went out for the members to form a line. Isnard did what he could to mimic the others, but he knew none of the society's ritualistic gestures, nor any of the songs, and with the society members clad in brilliant red-and-black robes he stood out like a sore thumb. In his own words he had yet "to take off his skin to put on the other."

The drums began, and the singing rose. The tension around him built terribly until a horrible face running with tears and blood cried out the lyrics that Isnard knew were meant for him: "That big, big goat in the middle of our house, the smart one is the one I want to catch." Our friend held his breath until his lungs hurt. He thought he was lost, and it was then that his genie came, not exactly possessing him but giving him the strength and mystical agility he needed. Just before they threw the first trap, the lights died and Isnard leapt out of line. They missed, striking instead the one standing next to where he had been. They tried again, and then again, until no fewer than ten members were caught by the trap. When the society leaders finally realized that this young man could not be caught, they sent three officials—the first queen, second queen, and the flag queen—to arrest him. Blindfolded, Isnard was taken before the cross of Baron Samedi to plead his case. Mercifully, the baron acknowledged his innocence, for at that very instant the song came into his mouth:

Cross of Jubilee, Cross of Jubilee
I am innocent!

Impressed by such an endorsement, the Bizango leaders took immediate steps to make Isnard a member of the society, and that night his initiation began. They taught him what he needed to know, so that now he could walk into any society gathering in the land.

Once initiated, however, Isnard discovered that the Bizango was unlike anything he had been told as a child. Rather than something evil, he found it a place of security and support. Whereas his mother had described the nocturnal forays of the societies as criminal and predatory, Isnard came to realize that the victims taken at night are not the innocent, but those who have done something wrong. As he put it, "In my village you kill yourself. People don't kill you." Those who must be out on the streets after midnight and who happen to encounter a Bizango band need only kneel in respect and cover their heads and eyes to be left alone.

Isnard also learned that he could appeal to the society in time of need. Should a sudden illness afflict a member of his family, the society would lend money to cover the medical bills. A member who got in trouble with the police, if innocent, could count on the Bizango leaders to use their contacts to set him free. Perhaps most importantly, Isnard found that the society could protect him from the capricious actions of his enemies. If, for example, someone should spread a rumor that cost you your job, by the code of the society you had the right to seek retribution. Again in Isnard's own words, "If your mouth stops me from living, if you oppose yourself to my eating, I oppose myself to your living." To exercise that right, a member need only contact the emperor—the founding president—and offer to "sell" the enemy to the society. The emperor, if he believed the case warranted a judgment, would dispatch an escort to bring both the plaintiff and the accused before the society. From what Isnard explained, however, it was not the flesh of the two that was presented but rather their ti bon ange, and though the experience would be remembered as a dream, their physical bodies would never have left their beds. This magical feat was accomplished by the escort—not a man, but the mystical force of the Bizango society. It cast a spell that caused the two adversaries to fall ill, and then just as death came near, it took the ti bon ange of each one. If your force was strong, if innocence was upon you, death was not possible, but the ti bon ange that was judged guilty would never return, and the corps cadavre of the individual would be discovered the next day in bed with the string of life cut. Selling an enemy to the society, however, was never done casually, for if the accused was deemed innocent at the tribunal, it would be the plaintiff who would be guilty, guilty of spreading a falsehood, and it would be he who was punished.

These revelations of Isnard, particularly the notion of "selling someone to the society," brought together the two separate but obviously related sides of the mystery. On the one hand there was the case of Clairvius Narcisse—his reference to "the masters of the land," a secret tribunal that had judged him, and a powder that had allowed him to pass through the earth. On the other stood the Bizango and their provocative but tentative link to the secret societies of West Africa, their knowledge of poisons, their use of tribunals and judicial process, their pervasive influence on community life. Our conversations with Isnard cast an image of the Bizango quite different from the popular stereotype, and at odds with Herard's dark picture of a wholly malevolent organization. Just how much of what Isnard told us was true we had no way of telling, but given his longtime friendship with the Beauvoirs we were encouraged to pursue our quest with him. As a youth Isnard had lived for some time with Max and his family, so we had this connection as a key to his confidence. Over the next ten days or so we met frequently both at his home and in the privacy of the Peristyle de Mariani. Our relationship with him blossomed, and he had managed to get us invited to a Bizango ceremony to take place the following week in Archaie when we received a summons from Herard Simon.

Herard wasn't on the corner by the theater as expected, and so while we waited Rachel and I took in the last few minutes of that week's movie, which turned out to be an unimaginably poor print of *Raiders of the Lost Ark.* The soundtrack was unintelligible, and as a result the movie became as much as anything a Rorschach test measuring the sensibilities of the audience. The climactic scene when the spirits shoot out of the ark and the flesh of the Nazi melts down was simply too much for many of the viewers. Pandemonium gripped the theater. Amid shouts of "Loup garou"—the werewolf—someone screamed a warning to pregnant women, and another cautioned all of us to tie ribbons around our left arm. It was a scene beyond anything in the picture itself. As the film ended the madness poured onto the street, and amid the shouts and laughter we barely heard Herard's harsh whistle. He had been in the theater the entire time, had in fact loved the movie, particularly the moment when the hero was trapped with the snakes in the Egyptian crypt.

"Someone born with a serpent's blood could do it," he assured us. "Otherwise it had to be a mystical thing." Herard explained that if you emptied your mind of all worries and made space you could shelter the spiritual allies that might allow a man to do the sorts of things that went on in the film.

"Only a fool," he added pointedly, his lips parting in the faint semblance of a smile, "would attempt to dance alone in the jaw of a lion."

Herard, of course, was aware of our recent activities in Archaie, our conversations with Isnard, even the invitation we had received to attend a Bizango ceremony the following week. I stepped away from his gratuitous, thinly veiled advice to join a small knot of moviegoers relieving themselves against the whitewashed wall of the theater.

"What kind of *blanc* pisses in the alley?" Herard called out as I came back to his crumbling jeep. "*Mes amis*, I can see it now. Once again my house is to become a resort of malfacteurs!" Trailing that laugh of his behind him, Herard stepped out of the small circle of light coming from the theater and without word or gesture led us away.

He carried a wooden sword as a staff, and his unwieldy pace took us from the center of town into a maze of narrow paths bordered by small houses of caked mud. It was getting late now, too late for most Haitians to be out. There was little movement, and with the blackout no light save that of the moon and the fitful glow of the odd lamp carelessly left burning. But the maze was alive with sounds—soft voices, babies crying, and the creaking of gates broken at the hinges. As we walked along, my imagination probed the darkness trying to pick out the meaning in the lives of these people living beneath thatch, surviving on the produce of gardens covered by crusted earth. From overhead came the slow drift of sweet ocean wind, sibilant among the fronds of the palm trees, and from the path the profane scent of man—squalid waste and rotting fruit, the corpse of a mule quivering with rats.

"*Honneur!*"

Herard had paused before a tall gate and rapped three times with his staff. No answer. He knocked again, repeating the customary salutation. There was some movement behind the gate but still no response. Finally, a lone woman's voice came out to interrogate the darkness. Herard named a man living on the other side of the compound and instructed her to go get him. When she refused, Herard's tongue lashed out, and the quiet alleyway exploded with all the intensity of an unbalanced dogfight. "Woman, guard your mouth!" Herard bellowed. "Do you want a coup l'aire? Shall I enter and shut your teeth? Shall I sow salt in every crease in your decayed skin?"

Another voice suddenly, and then the stiff click of the steel latch and the livid face of an old man, wild and ragged, poked out past the edge of the gate. When the door swung open and the woman saw whom she had been yelling at, she could not have appeared more frightened if confronted by a viper.

Herard gently ignored her remorse and motioned us to follow him through the tall gate into a compound set about with many huts. The living quarters were cloaked in darkness, but to one side separated by a small planting of bananas was a large *tonnelle*, the thatched canopy of a temple. Between the slats of bamboo that walled the tem-

ple we could see the flicker of lamplights, and numerous people passing silently before them. A man sat alone on a stool at the door of the
enclosure, his hands clasping a tin cup. Just as we stepped beyond the
flapping leaves of the plantains, whistles pierced the black air, and
from the shadows a small group of men appeared, stepped several
paces toward us, and then, seeing Herard, greeted him politely before
slipping away.

More than two dozen faces met us as we stepped across the threshold. At first obviously startled, within moments they had fallen back
within themselves, feigning a polite indifference. One row of benches
and another of cane-backed chairs stretched along a wall of the enclosure, and three vacant places appeared directly before the poteau mitan
in the front row of chairs. Herard told us to sit, then slipped out the
back door of the tonnelle. Within moments a matronly woman trailing
the sweet pungent scent of a Haitian kitchen—bay and basil, pepper-
corns and peanut oil—appeared with thimbles of thick syrupy coffee.
Across from us two young men were setting up a battery of drums;
they cast furtive glances at Rachel as she cracked open a bottle of
rum, tipped it three times toward the poteau mitan, and took a drink.
A murmur of approval ran along the benches behind her.

The peristyle was similar to others we had visited, with a single
centerpost and three sides of bamboo and thatch running upon a solid
wall of wattle and daub. A tin roof on top of flimsy rafters seemingly
supported by the web of strings displayed what must have been a hundred faces of President Jean-Claude Duvalier and ten times as many
small Haitian flags. Three doors in the wall led to the inner sanctum of
the temple, and there were two exits—the doorway we had entered
through, and an open passage at the other end through which Herard
had left. The benches were full, but people continued to arrive—the
men stolid and unusually grave, the women determined, and everyone
wearing fresh clothes: cotton beaten over river rocks, hung out to dry
on tree limbs, pressed with ember irons and liberally scented with tal-
cum powder. As at other vodoun gatherings I had attended, there
were more women than men. Many of them arrived with pots of food
that were hastily carried to charcoal cooking fires flickering beyond
the far end of the enclosure.

But there were other symbols before us I had never seen: the
black skirting around the pedestal of the poteau mitan; and the color
red in the cloth that entwined the centerpost, on the wooden doors

that led into the bagi, and in the flags whose significance I finally understood. Red and black—the colors of the revolution: the white band in the French tricolor ripped away and the blue in the time of François Duvalier becoming black, the color of the night, with red the symbol of transition, lifeblood, and rebellion. On the wall hung paintings of the djab—the devil—gargoylelike with protruding red tongues pierced by red daggers and lightning bolts, and with the inscriptions beneath, again in red paint dripping across the whitewash, "The Danger of the Mouth." There were figures of other strange spirits—Erzulie Dantor, the Black Virgin; and Baron Samedi, the Guardian of the Cemetery. Arched across one of the entrances to the bagi, again in red but this time etched carefully in Gothic script, was yet another inscription, which read "Order and Respect of the Night." Finally, at the foot of the poteau mitan lay a human skull wearing a wig of melted wax and crowned with a single burning candle. Herard Simon, defying all our expectations, had brought us to a gathering of the Bizango.

Rachel's hand touched mine, and her eyes led me toward an imposing figure standing alone by the rear exit, beckoning us to his side. We followed him outside, past the cooking hearths and across a small courtyard, until we reached Herard and another man sitting together on the lip of a well.

"So," Herard began, "you are where you want to be. Salute the president and ask what you will."

It had happened that quickly. Stunned, we groped for words in the darkness. Herard leaned back, both hands resting on the hilt of his wooden sword. The president lifted his face to us and spoke in soft, velvety words. "*Mes amis*, has the djab seized your tongues?" His tone was gentle and surprisingly kind.

"It is Monsieur . . . ?" Rachel began.

"Jean Baptiste, mademoiselle, by your grace."

"It is my father, Max Beauvoir, who brings us to you. He serves at Mariani. Perhaps you have heard him on the radio?"

"I know your family, Rachel. I am of Saint Marc, and your uncle will tell you of me. But what is it that has led you to me tonight? Your father has shared his table with me, but now sends his daughter as an envoy to the Shanpwel?"

"There are things a child of Guinée must understand," Rachel said. She began a brief summary of our previous activities, but was abruptly cut off by Herard.

"These things are well known. He doesn't have all night."

"Women," she went on hesitantly, "have their place in the Bizango?"

Jean Baptiste didn't answer. His eyes looked beyond us, sweeping the compound.

"Rachel," Herard interceded again, "are you not the daughter of your father? Women have a function everywhere. What kind of society wouldn't have women?"

"The people say," she tried again, more boldly, "that the society can be as sweet as honey or as bitter as bile."

The president lit up. "Ah, my friend," he sighed, "she is already a queen! Yes, this is so. Bizango is sweet because it is your support. As long as you are in the society you are respected and your fields are protected. Only the Protestants will hate you."

"And bitter?"

"Because it can be very, very severe." The president paused and glanced over toward Herard, who nodded in affirmation. "Bizango is a great religion of the night. 'Order and Respect of the Night'—that is the motto, and the words speak the truth. Order because the Bizango maintains order. Respect of the night? As a child, Rachel, what did your father teach you? That the night is not your own. It is not your time, and you must not encounter the Shanpwel because only something terrible can occur. Night belongs to the djab, and the eyes of the innocent must not fall upon it. Darkness is the refuge of thieves and evildoers, not the children. It is not a child that should be judged." The president looked up. "It is time," he said softly, excusing himself. "Stay with us now and dance. Tomorrow you will visit me at my home in Saint Marc."

The drums began, and as we passed back into the tonnelle small groups of dancers swirled before them, challenging each other to greater and greater bursts of effort. The appearance of Jean Baptiste had a sobering effect, and as he began to sing the dancers drew into a fluid line that undulated around the poteau mitan. Knees bending slightly and bodies inclined backward, they answered him with a chorus that praised Legba, the spirit of the crossroads. Salutations to other familiar loa followed—Carrefour, Grans Bwa, Aizan, and Sobo. It struck me that this gathering of the Bizango was like the beginning of any vodoun service. Women continued to mingle about, and the old men on the cane-bottomed chairs passed bottles of clairin peppered with roots and herbs. An hour went by, and although the atmosphere became charged with activity no spirit arrived. Instead, just before eleven o'clock, a cry went out from the president and was answered by all present.

"Those who belong!"

"Come in!"

"Those who don't belong!"

"Go away!"

"The Shanpwel is taking to the streets! Those who belong come in, those who don't go away! *Bête Sereine!* Animals of the Night! Change skins!" A man with ropes across each shoulder like bandoliers came hurriedly across the tonnelle carrying a sisal whip and took up a position just outside the exit. The door slammed shut, and the dancers rushed into the bagi.

"Seven cannon blows!"

From outside came the deep penetrating moan of a conch trumpet, followed by seven cracks of the *fwet kash*, the whip. Moments later the dancers reemerged from the bagi and, encouraged by steel whistles, took up marshal positions around the poteau mitan. On one side stood the men, dressed uniformly in red and black, and across from them the women, clad in long red robes.

With the president standing to one side and all the members in place, a woman with a high, plaintive voice sang out a solemn greeting asking God to salute in turn each officer of the society. As they were called, one by one men and women stepped out of rank, stood before the group, and then, following the slow, almost mournful rhythm of the prayer, moved to form a new line still facing the poteau mitan. Their titles were mostly unfamiliar to me: *secrétaire, trésorier, brigadier, exécutif, superintendant, première reine, deuxième reine, troisième reine.* Suddenly a whistle broke the tension, eliciting shrill ritual laughter from the ranks. Unaccompanied by the drums, and with the members formally bowing and curtsying in time, the society began to sing:

> I serve good, I serve bad,
> We serve good, we serve bad,
> Wayo-oh!
> When I am troubled, I will call the spirit
>   against them.

Then came a song of warning, set off from the last by whistles and the cracking of the whip:

What we see here
I won't talk,
If we talk,
We'll swallow our tongue.

The singing continued until three people reemerged unexpectedly from the second door of the temple. One was the secretary, only now he carried a machete and a candle and wore a black hat covered by a sequined havelock. Beside him came a woman, perhaps one of the queens, cloaked in green and red, and following these two a woman in red with a small black coffin balanced on her head. The rest of the society fell in behind her, and singing a haunting hymn of adoration, the small procession circled the poteau mitan, eventually coming to rest with the woman laying the coffin gently on a square of red cloth. An order went out commanding the members to form a line, and then the secretary, the president, and one of the women walked ceremoniously to the end of the tonnelle, and pivoting in a disciplined military manner returned as one rank to the coffin.

President Jean Baptiste, flanked by his two aides and speaking formally in French, officially brought the assembly to order.

"Before God the Father, God the Son, God the Holy Ghost, I declare this séance open. Secretary, make your statement!"

Holding his machete in one hand and a worn notebook in the other, the secretary exclaimed, "Gonaives, twenty-four March nineteen eighty-four, Séance Ordinaire. By all the power of the Great God Jehovah and the Gods of the Earth, and by the power of the Diabolic, of Maître Sarazin, and by authority of all the imaginary lines, we declare that the flags are open! And now we have the privilege of passing the mallet to the president for the announcement of the opening of this celebration. Light the candles!"

One by one each member solemnly stepped out of line and in graceful gestures of obeisance paid homage to the coffin, leaving a small offering of money and taking a candle, which was passed before the flame burning at the base of the poteau mitan. A line of soft glowing light spread slowly down one side of the tonnelle, and the society members, their heads bowed and their hands clasping the candles to their chests, began to sing once more. Three songs, each an eloquent call for solidarity. The first posed a potent question:

President, they say you're solid,
And in this lakou there is magic.

> *When they take the power to go and use it outside,*
> *When they take the power who will we call?*
> *When we will be drowning, what branch will we hold*
>   *onto?*
> *The day we will be drowning, who will we call?*
> *What branch will we hold onto?*

Then, as if to answer this lament, the second song continued:

> *Nothing, nothing can affect us,*
> *Before our president,*
> *Nothing can hurt us,*
> *If we are hungry, we are hungry among ourselves,*
> *If we are naked, we are naked among ourselves.*
> *Before our president nothing can harm us.*

The final song was raised in defiant, raucous voices with the feet of all members stamping the dry earth, raising small clouds of dust:

> *I refuse to die for these people,*
> *This money is for the djab,*
> *Rather than die for people, I'd rather the djab*
>   *eat me,*
> *This money is for the djab,*
> *I will not die for these people!*

By now the tension in the rank had become palpable, and the president had to slap a machete against the concrete base of the poteau mitan to restore order. At his command a man later identified to us as the treasurer advanced with one other to count the money, and with an official air they announced, "Sixteen medalles—sixteen gourdes"—a sum of less than five dollars. The president stepped forward reciting a Catholic litany blessing the offering and seeking the protection of God for the actions of the society. As his final words expired, the drums exploded, breaking open the ranks at long last. Other songs rose in strange rhythms, with one of the four drums played lying on its side, giving the staccato sound of wood striking hollow wood. The drums ceased as suddenly as they began, and once again attention focused on the president, now standing alone by the poteau mitan, his hands cradling a weeping and terrified baby. His own voice, high and soft with reverence, was soon joined by all the others:

> *They throw a trap to catch the fish,*
> *What a tragedy! It is the little one who is*
> * caught in the trap!*

The baby's mother stood by the president's side, and as the others sang tears came into her eyes and ran in rivulets down her cheeks. With gentle gestures Jean Baptiste led her and the baby around the poteau mitan, and then he lifted the child tenderly above his head to salute the four corners, the four faces of the world. As he turned slowly, the society members beseeched him:

> *Save this little one,*
> *Oh! President of the Shanpwel!*
> *Save the life of the child we are asking,*
> *Oh! President of the Shanpwel!*
> *Save the life of this child!*
> *Yawé! Yawé!*

The unbroken circle of the Shanpwel closed around the body of the president, and one by one they lifted the child from his arms and bathed it with a warm potion of herbs. Then, the treatment completed, the drums resounded once more, and the members of the Bizango danced long into the night.

Sunlight has a way of diffusing mysteries, and the next afternoon, as Rachel and I sat near the waterfront of Saint Marc waiting for Jean Baptiste, all we could think about was the heat and the surge of flies hovering about us. It was as if all the lurid tales of the Bizango had given way to this: a shoreline and the smell of fish, briny nets and cracked tar rising to mingle with the dust of the city. No babies had been slaughtered the night before, nobody had been transformed into a pig, least of all the two of us, who rather had been treated graciously as honored guests. Quite contrary to the image I had been given, the gathering of the Bizango had impressed me as a solemn, even pious, ceremony that revealed among other things a strict hierarchical organization modeled at least superficially on roles derived from French military and civil government. Whether this was a purely symbolic hierarchy or something more remained to be seen.

We had spent much of the morning trying to find out more about Jean Baptiste, speaking with Rachel's uncle Robert Erié, at one time the prefect for Saint Marc. Mr. Erié is a kind, generous man, and though a large landowner and leader among the local bourgeoisie, he is clearly respected and liked by the entire community. Though unaware of Jean's position as president, Robert Erié knew the man well. He had in fact employed Jean as a chauffeur for the entire time he had been prefect. A diligent and competent driver, Jean was remembered as being a good man, quiet and discreet, but not particularly influential in the traditional peasant society of Saint Marc. Although I said nothing about it, it had struck me as significant that the president of a secret society should be employed in such a capacity. During the colonial era many of the slaves who eventually became important leaders of the revolution had worked as coachmen, an ideal position from which they could spy on the ruling authorities.

Speaking with the former prefect had offered an unusual insight into how a prominent official appointed by the central government interacts with the traditional society. Despite his ignorance of Jean's position, he knew about the activities of the Bizango in some detail. As prefect, he explained, it had been his responsibility to know what went on in the region under his jurisdiction. As an official he didn't condemn or even judge what the societies did. They were his friends, and he moved freely among them, "drinking his drink," as he put it, "without problems." But they in turn had the responsibility of keeping him informed. No ceremony, for example, should be held without the prefect's preknowledge. Erié considered this a matter of courtesy, not coercion, and stressed that during his entire time in office he had not once taken steps to interfere. Nor, he added, could he imagine a situation when this would be necessary. This easy relationship between the representative of the urban-based authorities on the one hand and the traditional leaders of the vodoun society on the other is no accident, I was to learn. On the contrary, it is mandated by the very nature of the contemporary government in Haiti.

The civil government of Haiti is divided into five *départements*, and each of these, in turn, is divided into a number of *arrondissements*, with each one headed by a prefect appointed directly by the president of the nation. Each arrondissement—there are twenty-seven in all—is composed of *communes* led by the equivalent of a mayor assisted by an administrative council that is based in a village. Beyond the edge of the village itself, the land of the commune is divided into a number of *sections rurales*.

The military have their own parallel subdivision of the country, and while it is somewhat different, the important point is that at the lowest level the two systems come together, making the rural section the basic level of local government. It is within these rural sections that at least 80 percent of the Haitian population lives.

There is a curious and important paradox in the governing of these rural sections, however, and it hinges on the role of an official outside the hierarchical organization of either form of government. Gerry Murray, one of the most thoughtful anthropologists to have worked recently in Haiti, has pointed out that the rural section in no way coincides to a community or village, but rather describes "an arbitrary administrative lumping of many communities for the purposes of governance." The rural peasants themselves identify not with their section but with their own extended families and neighbors in its lakous, the familiar compounds made up of clusters of thatched houses one sees all over the country. In other words, neither institution of the government, the civil or the military, recognizes in any juridical sense the actual communities in which the vast majority of the rural peasants live and die. To reach these people the national authorities depend on one man, the *chef de section*, an appointee from *within* the rural sections who is expected to establish networks of contacts that will place his eyes and ears into every lakou in his jurisdiction. This he does, but in a very special way.

Although the chef de section derives his *authority* from the central government, the basis of his *power* is not his official status so much as the consensus of the residents of his own section. He does not act alone, but rather heads a large nonuniformed force of local peasants that in turn derive their extrastatutory authority not from him but again from their own people. The chef de section can be quite helpless without such popular support, and historically efforts of the central government to place outsiders in the position—most notably when the American occupation forces attempted to replace vodounists with literate Protestants—have always failed. Haitian law provides for the chef de section to retain certain assistants, but as Murray indicates, "the particular form the structure of police control will take in a particular region will be largely governed by local traditions and by adaptive adjustments to local social reality on the part of local law enforcement officials, who are themselves intimately familiar with this reality." What this implies is that the official government, in order to reach the peasants, must tap into their own traditional networks of social control, in the person of the chef de section.

Who is this chef de section? Above all he is himself a local peasant. Typically he maintains his own fields, is polygynous, and serves the loa. In many instances he is a prominent houngan. His behavior, personal values, and expectations are not those of a bureaucrat, but rather those of a leader of the traditional society. Although technically he receives a salary from the capital (and sometimes the money actually arrives), he depends financially on his own land, and like the African patriarch he considers it his right to recruit unpaid labor for his fields. This service the community members willingly provide, in effect as compensation for the hours he must spend attending to their affairs. It is his task, after all, to investigate conflicts and convene the informal tribunals at which virtually all local disputes are said to be resolved. With their power thus rooted in their own jurisdictions, the chefs de section remain relatively unaffected by political upheavals in the capital, and as Murray points out they retain, sometimes for indefinite periods, virtual unchallenged control of their areas. Threats to that power and their position, then, come not from Port-au-Prince but from dissatisfaction among their own people, so it is with them that the loyalties of the chef de section lie.

In brief, though the institution of the chef de section serves as an interface between the two separate worlds that make up the Haitian reality, the man himself is a member of the traditional society, and the network of contacts he taps is the network within that society. It was, therefore, by no means a trivial discovery to learn from Rachel's uncle Robert Erié that in most instances the chef de section was also the president of a Bizango secret society.

Jean Baptiste surprised us by turning up dressed in the uniform of an army corporal and driving a military vehicle, looking like all the anonymous soldiers one runs into at the scattered roadside outposts in Haiti, the ones you catch napping behind their desks when their subordinates lift you from your car for a passport check. Yet within moments of stepping off the street into the privacy of his home he again assumed the air of quiet dignity and authority that had so impressed me the night before. He spoke freely, answering every question we put to him, and as we listened the nefarious facade of the Bizango crumbled piece by piece.

"No, no. It was an accident. Children are the little angels. They can do no wrong. Children do not fall within the sanction of the society. What you saw was an unfortunate victim. Another society needed to take someone, but the coup l'aire fell upon her by mistake."

"Who did they want to take?"

"That is their business."

"It could be anyone?"

"Not at all. Listen, you must understand that the Bizango is just like a normal government. Everyone has their place. It is a justice." Jean Baptiste reviewed each member's grade. The leadership consisted of the emperor, presidents, first, second, and third queens, chef détente, and vice-president. This ruling body was known as the chef d'état-majeur. There were other queens—the *reine dirageur*, who was generally the emperor's wife; the flag queens; and the flying queens. Other lesser positions read like a list from the French civil service—prime minister, conseiller (advisor), avocat (lawyer), secrétaire, trésorier, superviseur, superintendant, intendant, moniteur, exécutif. Military titles included general, prince, brigadier general, major, chef détente and soldat. Finally, there were three positions of note: the bourreau, or executioner, whose task it was to enforce the collective decisions of the society; the chasseur, or hunter, who was dispatched to bring culprits before the society; and the sentinelle, who was positioned at the gate as a scout to prevent unwelcome individuals from entering a society gathering.

According to Jean Baptiste, his was but one of as many as a dozen Bizango societies in the Saint Marc region alone. Each one maintained control over a specific area and was led by a founding president known also as the emperor. In time members of the society's hierarchy might launch out on their own, establishing their own bands still within the umbrella of the original leader, and as a result within any one Bizango territory in Saint Marc there might be as many as three societies operating. Each principal society was supposed to respect the borders of its neighbors, and disputes if necessary could be taken before the one selected by all the society emperors to reign as the emperor of the entire region.

The purpose of the Bizango, as Jean Baptiste insisted on reiterating, was to maintain order and respect for the night. To its members and their families it offered protection, and since almost everyone had some relative initiated, the Bizango became the shelter of all.

"For the sick, or the troubled," he explained, "it is a wonderful thing, whether you have money or not. But it is only good because it can be very bad. If you get into trouble with it, it can be very hard on you."

"Do you mean someone bothering the society, or just anyone?"

"No, of course not anyone. I tell you it is a justice, so they must judge you. Listen, say someone is out on the street and the time is not his. Then perhaps he will receive a coup l'aire just as a warning. He becomes sick a little, and then he remembers not to walk by night." Jean shrugged, as if what he was saying was the most obvious thing in the world. "But if someone disturbs a member of the society for no reason it will be something else. The guilty one will be sent to sit down. You see, we have our weapons."

That arsenal, as Jean explained, contained a number of spells and powders—the familiar coup l'aire, coup n'âme, and coup poudre—that were surreptitiously applied to the victim at night. To do so the society executioner set traps in places the person was known to frequent. A powder sprinkled in the form of a cross on the ground to capture a ti bon ange was one example of such a trap. Once caught, the victim could only be saved by the intervention of the society president, a treatment that could cost him dearly.

"They say we eat people," Jean continued, "and we do, but only in the sense that we take their breath of life."

"But Herard said that the Bizango will do far more than that."

"The commandant says what he wants to say."

"He's a society of one," I said with a smile, remembering Herard's words at Petite Rivière de l'Artibonite. Jean looked at me, but he didn't return my smile.

Just what was going on was uncertain—as was whether Herard was still involved. The image he had presented so forcefully of the Bizango was unlike anything we were seeing with Jean Baptiste; yet at the same time he had led us to the ceremony, introduced Jean, and sat all evening delighted to see us participate.

"When might a society take someone?" Rachel asked.

"They are out at night," Jean began, then hesitated, "or word may come through even by day. Just as long as it becomes known that someone has been talking improperly. Perhaps he is the one who runs to the *blanc* or the upper-ups in the government and tells them what such and such a society is doing. The person may try to hide, but we shall find him. That is how a man ends up passing through the earth twice."

"A zombi?" I asked.

Jean regarded me steadily. "Every evildoer will be punished. It is as we sing." In a high voice that took us back to the night before at Gonaives, he proceeded to sing:

> *They kill the man*
> *to take his zombi*
> *to make it work.*
> *O Shanpwel O*
> *Don't yell Shanpwel O*
> *They killed the man*
> *to take his zombi*
> *to make it work.*

Jean was laughing aloud even before he finished the verse.

"And what about thieves?" Rachel asked.

"You need proof, you always need proof. If someone complains to the society, they will investigate. If it is true, the trap will be set. If a Shanpwel steals, he is rejected immediately from the society."

"And if he escapes?" I asked, recalling the way Isnard described dodging the trap.

"For those who are guilty there is no escape. The society will follow him to the ends of the land, even to the height of the Artibonite."

"Across the water?"

"Even to the streets of New York."

"But on the land, is this not the work of the chef de section?"

"Yes. It is all the same. The society moves by night, and since most crimes are done in the dark we naturally prevent many things. We work together. The chef de section must know what goes on, and so we tell him. He'll find out anyway. If the society is going into a territory to take someone, and the chef de section is told and he knows the culprit is guilty, he won't interfere. Anyway, most often the chef de section is a houngan or president Shanpwel. He walks by day, but by night he too changes his skin."

"The plate needs the spoon and the spoon needs the plate," Rachel said, echoing the words of Marcel Pierre.

Jean laughed once more. "My friends, what you have seen is so little." He explained that the ceremony the night before had been no more than a bimonthly training session of a single society. The next week there was planned a regional gathering of all the Saint Marc bands to celebrate the founding of a new society. It would be his honor if we would accompany him as his guests.

13

Sweet As Honey,
Bitter As Bile

BLOOD IN HAITI costs the peasant twenty dollars a liter, assuming he can get the official price at the door of the Red Cross blood bank. Marcel Pierre, Rachel, and I stood by the hospital bed and watched yet another bag of it enter the frail arm of Marcel's wife, and almost as quickly seep out between her legs to soak the cotton sheets. Public hospitals in Haiti have something in common with the jails; the care of the inmates depends less on their condition than on the ability of their kin to pay—for medicines, food, bedding, even rental space on the lumpy mattresses left over from the American occupation. Marcel had sold most of what he owned, and still no one had done anything to cure his wife. He had shuttled her from hospital to hospital, bringing her south from Saint Marc by *camionnette* squeezed in among the market women and chickens. Earlier in the day he had spent his last money getting her to the capital only to have to watch her now, lying in the hospital bed, knowing that if he didn't get more blood by nightfall she would die.

It was under these conditions that his pride allowed him to turn once again to us. For some time I had been giving him money—"advances," as we agreed, against another preparation. He knew I didn't need it, just as he knew that the money I had already paid him was far more than any powder was worth. Still, all this remained unsaid that evening when he turned up forlornly at the Peristyle de Mariani. Rachel kissed on him on both cheeks and rushed off to fetch a tray of food. Marcel and I embraced, then sat together on the couch. By now, of course, we knew all about him, not only his business and reputation, but also his relative position in the traditional hierarchy of Saint Marc. He was small stuff, really, a houngan *nieg* as some had said—one who must walk the streets to have any presence at all. Marcel was an outcast, a pimp, a malfacteur, a powderer, and now more than ever I sensed his isolation. Haitian men do not cry, but tears slipped from Marcel's eyes. My hand held his, but there was nothing more I could do. Watching Marcel beside me, I realized just how far he and I had come from the early days and the foolhardy images we had both projected to each other.

Max Beauvoir knew all about Marcel's reputation as well, but when he walked into the room with Rachel he received him as a peer. It was an important gesture, one that moved Marcel visibly. And that night before the ceremony, Max took Marcel aside, and they worked together in the inner sanctum of the temple. Later, after the opening prayer, he honored Marcel by inviting him to lead the ceremony. Marcel took this to heart, and the singing transformed him. In time the spirit Ogoun came up on him, and he raced around the periphery of the peristyle, ripping the tablecloths from beneath the drinks of the startled audience. Beauvoir did nothing to stop him. Ogoun burned with a raging intensity. An hour before, death had seemed so close. Now the man had become a god, and death was suddenly an impossibility.

Rachel and I spent most of the next day moving around Port-au-Prince with Marcel—buying more blood, finding a clinic for his wife, a rooming house for his daughter. It was one reason we were late getting to Saint Marc that night for Jean Baptiste's Bizango ceremony. There were others. The rain again flooded the Truman Boulevard, backing up traffic halfway down the Carrefour Road, and then we ran into the presidential motorcade that blocked the road just north of the capital. Finally, there was Isnard, who had turned up unexpectedly at

Mariani with an invitation to attend a Bizango ceremony early that same evening in Archaie. By the time we had negotiated the various traffic snarls and ferried Isnard on a number of last-minute errands in the capital, it was already well past dark. A blinding rain slowed our journey north, as did the unsuccessful diversion at Archaie. The site of Isnard's reputed ceremony was deserted and the hounfour shrouded in darkness, save for the warm embers of a recent fire. Behind a crack in the sealed door of a neighboring hut a lone woman dressed in red and black informed Isnard that someone had died during the ceremony, and the society had fled. By the time our jeep once more gathered momentum against the deepening night, we knew that we would be late for our rendezvous with Jean Baptiste.

Even Jean's oldest son, with his father already having left, did not know the exact location of the Bizango gathering, but following his father's instructions he led us along a fretted track to a crossroads just south of the town. There we waited. The storm had passed and the trade winds overhead blew away the last wild rage of tattered clouds revealing a night sky deep with stars. The stillness was broken only by the sound of the waves beating the shoreline. When, after close to an hour, the expected contact failed to appear, we swallowed our disappointment and began to make our way slowly back toward Saint Marc. It was just after we turned away from the sea that Isnard heard the sound of the Bizango drums. Leaving him by the roadside, we returned to town to drop off Jean's son, then immediately doubled back. By then Isnard had discovered the source of the distant rhythms. He led us to an open field enclosed by a living fence of caotchu. At the center of the field stood a complex of low buildings.

Instantly an enormous man bearing the rope bandoliers of the sentinelle stepped before us to plug the gap in the fence. A second guard appeared in the path behind, blocking our rear.

"Who are you?" demanded a harsh anonymous voice.

"Bête Sereine. Animals of the Night," Isnard replied flatly. The sentinelle moved aside, revealing yet a third Shanpwel, who accosted Isnard by clapping his hands on each of our friend's shoulders.

"Where are you coming from?" he asked.

"I come from the heel."

"Where are you going?"

"I'm going into the toes."

"What toe?"

"I am the fifth."

"How many stars do you walk with?"

"We are three." Isnard offered his hand but the gesture was ignored. The arms of the shadowy figure remained firmly anchored to his sides. Rachel moved confidently forward.

"We have come to you by invitation of our friend Jean Baptiste," she said politely. The man glanced at her suspiciously.

"He expects you and the *blanc?*"

"Yes, we are late. There was too much rain." Loosening his hold on Isnard, the man dispatched the sentinelle to take our message into the ceremony. For several awkward minutes we stood quietly, listening to the rhythms of the Bizango drums. Though the temple was but a hundred meters away, the sound was fainter than it had been when we first heard it from the roadside. It was just as Herard had said. "When the Bizango drum beats in front of your door, you hear it as from a great distance, and when it is miles away it sounds as if resonating in your own yard."

The arrival of a messenger from Jean Baptiste broke the tension. The man's warm greeting sent the sentinelle and his companions back into the shadows. Chatting idly, then falling silent, we followed the messenger along a narrow path toward the temple. The incessant drumming continued to soften, and as we drew near we heard the words of the song:

> *The band is out, the society is out,*
> *Watch out mother who made me.*
> *The band's out, the society is out*
> *Mothers of Children*
> *Tie up your stomachs.*

The society was out, its procession winding past some distant crossroads, but the temple was still packed, a catacomb of gestures and faces that seemed vaguely inimical as we eased our way into the crowded entry. To one side in a small open space around the poteau mitan a handful of men and women dipped and swayed, and just beyond, hidden from the light, a group of drummers stood straddling their instruments, teasing the dancers like puppeteers. The music didn't stop, but the drummers looked up, and their throaty irreverent voices announced our presence in no uncertain terms.

The messenger led us out the back of the temple into a courtyard vibrating with several hundred people. There was a festive air, and all the crowded tables together with the lingering odor of grilled meat and sweet potatoes reminded me at first of a community picnic. But the atmosphere was also charged with tension, somehow accentuated by the loud hissing of dozens of Coleman lanterns set around the periphery of the courtyard. They cast a harsh, almost blinding light that kept my eyes to the ground as Rachel, Isnard, and I wove past a number of tables until we reached the one at the back headed by Jean Baptiste. We were welcomed and invited to sit by his side, but there were no introductions, and when Rachel finished apologizing for our delay and exhausted a few easy phrases of greeting, a disconcerting stillness gripped us. Across the table a half-dozen men sat as still as stones. The light from behind them completely obscured their faces in shadow and reduced their enormous bodies to a single ominous silhouette. I started any number of stories that had always worked in the past but now drew no response. Isnard was equally uncomfortable, yet our presence challenged his pride to display his knowledge of the Bizango, and in a foolish gesture he pulled a demijohn of clairin from his hip pocket and stood up to propose a toast. To make matters worse, the bottle was empty.

"Friends of the night," he said with inappropriate melodrama, "we are all brothers."

A flat voice responded from across the table. "What do you mean all brothers?"

"Well," Isnard stumbled, "we are Haitians, we are all Haitians." He had intended, of course, to imply that he too was a member of the Bizango.

"Yes, it is so. We are *Nég Guinée*, the people of Africa," the shadow agreed. "Some of us."

Fortunately, before Isnard could go any further, the tables around us began to buzz with the rumor of the returning procession, and within moments the entire courtyard was filled with anticipation. Jean led us quickly into the peristyle, which, already crowded when we arrived, was by now bulging with members of the Bizango. In the distance, seen through the gaps in the bamboo wall, the procession like an articulated serpent crawled over the hills—proceeding, halting, then moving again, slowly emerging out of the nether darkness of the night. Lanterns on either flank floodlit great swirls of chalky dust. You could hear the distant voices, the whistles, and the short barking commands driving the members into line. The crack of sisal whips kept time with the ponderous cadence of the march as the leaders paused to allow the coffin to sway like a pendulum hanging from the shoulders of the escort. To this fusion of red and black, the droning voices lent a certain uniformity binding everyone present to the rhythms of the music.

> *We come from the cemetery,*
> *We went to get our mother,*
> *Hello, mother the Virgin*
> *We are your children*
> *We come to ask your help*
> *You should give us courage.*

As the procession poured into the temple, clusters of lanterns at the entryway outlined the stern faces of the magnificently robed acolytes. First inside was the sentinelle, who leapt around the poteau mitan in a mock display of suspicious concern. Behind him walked a lone man carrying a rope, followed closely by the sacred coffin and its circle of retainers. As the bulk of the society squeezed into the room, the guests pulled back into an adjoining chamber until the enclosure swelled with red-and-black robes. Slowly the procession circled the poteau mitan, but there were so many that the members didn't walk so much as lean into each other, flesh to flesh, sending a single wave surging around the post. A tall, dazzling woman wearing a satin and spangled Edwardian dress and a buccaneer's hat complete with ostrich plumes took the center of the room. With slow sweeping gestures she orchestrated the entire movement while her face, half hidden beneath her hat, radiated perfect serenity, a look both world-weary and utterly calm.

The power of the moment was dreamlike. The sacred coffin rested on the red-and-black flag, caressed by flowers. The Shanpwel blew out the burning candles they had clasped to their chests and laid them in one of two velvet hats placed at the foot of the coffin. The majestic woman in the glittering robe moved at ease through the huddled members of her procession, nodding to each one. Whistles and conch shells echoed as a great shout arose.

"Twenty-one shots of the cannon in honor of the executive president!" As the whip outside began to crack, the woman, her head and great hat tilting slightly like the axis of a globe, turned slowly back toward the base of the poteau mitan. She was the empress, I now realized, and we were all gathered to celebrate the founding of her society. A man acting as a herald stepped forthrightly to her side.

"Silence! In a moment we . . ." he began, but the raucous excitement kept him from being heard. "Silence! People, you are not in your houses! Be quiet!" The empress raised an arm and the room fell silent.

"In a moment," the man continued, "we shall have the honor of presenting you with the attending presidents." Then one by one as their names were called five Bizango leaders including Jean Baptiste stepped forward. The empress, lifting her face to the light to reveal the finest of features—thin bones and skin drawn beautifully across high cheekbones—formally opened the ceremony.

"Circonstantier, thirty-one March nineteen eighty-four, Séance Ordinaire. By the entire power of the great God Jehovah, Master of the Earth!"

A short man moved forward to assume the center of attention. As his dazed eyes scanned the room, silence followed easily.

"My dear friends," he began. "Often on a day such as this you have seen me stand before you, and for me it is a joyous occasion, for it allows a chance to welcome you and to share through my words a number of ideas and themes that mean so much to us all. Sadly, tonight my thoughts have escaped, my imagination has escaped, and this prevents me from yielding to my strongest desires. Still, I wish to thank all the guests for responding to the invitation of this morning."

With this introduction he fell quiet for some time. It was as if for him, with his oral tradition, the spoken words were alive and each had to be savored.

"Brothers and sisters!" he continued. "It is today that we come together in a brotherly communion of thoughts and feelings. Feelings that are crystallized into the force of our brotherhood, the force that gives meaning to our feast of this night.

"Yes, we are talking about our Bizango institution, in the popular language Bizango or Bissago. Bizango is the culture of the people, a culture attached to our past, just as letters and science have their place in the civilization of the elite. Just as all peoples and all races have a history, Bizango has an image of the past, an image taken from an epoch that came before. It is the aspect of our national soul."

Nobody made a sound as he paused again, and when he resumed, his words spread over the cushion of moist hot bodies, mingling with the heat of the lanterns until the air was vibrant with the ebb and flow of a timeless idea.

"The Bizango brings joy, it brings peace. To the Haitian people Bizango is a religion for the masses because it throws away our regrets, our worries, problems, and difficulties.

"Another meaning of the Bizango is the meaning of the great ceremony at Bwa Caiman. They fall within the same empire of thoughts. Our history, such moments, the history of Macandal, of Romaine La Prophétesse, of Boukman, of Pedro. Those people bore many sacrifices in their breasts. They were alive and they believed! We may also speak of a certain Hyacinthe who as the cannon fired upon him showed no fear, proving to his people that the cannon were water. And what of Macandal! The one who was tied to the execution pole with the bullets ready to smash him but found a way to escape because of the sacrifice he did. Again we see . . ."

Suddenly a harsh voice from across the room interrupted the speech, directing the gaze of all present at Rachel and me. I recognized the rotund man as one of the presidents who had been introduced earlier.

"By your grace, my friends. Dessalines used to tell us things. I am a man. I am not in my house, but I can talk. Why is it that a *blanc* is here to listen to our words?" A shiver of confrontation ran through the people around us. With the permission of Jean Baptiste I had been recording parts of the ceremony openly, and now my tape became the center of concern.

"Are our words meant to pass beyond the walls of this room?" In the stiff silence I could hear raindrops hissing as they dropped into a small flame by my ear.

"Give me the tape, quickly," Rachel said. I did so, and stepping before the assembly she offered the cassette to the offended party and begged his forgiveness. "The *blanc* is here as my guest, and I am here because I am a child of Guinée." The tape was accepted with equal grace, and the crisis seemed to have passed when Jean Baptiste stepped forward, face to face with the other president.

"The tape is mine. The *blanc* is my guest also. My house is his, and the girl is the child of my friend." Thus, after a moment's hesitation, the tape came back to me. An uneasy calm returned to the room, and the intense man standing by the poteau mitan resumed his speech from the beginning, closing with a pointed reference to the Haitian revolution.

"We say again that there was a moment in eighteen-oh-four, a moment that bore fruits, that the year eighteen-oh-four bore the children of today. Thank you."

In a prudent attempt to reestablish the rhythm of the celebration, the empress called immediately for the offerings to the sacred coffin. She led a prayer and then sang the opening verses of the adoration hymn, as one by one each person in the room came forward to place a small sum before the society.

"You certainly won't go through this if you visit my society." A short, elfin woman had nestled up beside me, and was speaking softly. "This is crazy. It's a public ceremony. Everyone should be welcome." Before we knew it the woman was standing between Rachel and me gently holding each of us by the hand. Her name was Josephine, and her society met every second Wednesday. We were more than welcome to attend, she assured us. A nattering, affectionate conversation ensued between her and Rachel, until interrupted by an aggressive, threatening voice coming from the other side of the room.

"Brothers and sisters! Wait a minute! The ceremony at Bwa Caiman was a purely African affair. There should be no *blanc* assisting with this ceremony of ours tonight! No *blanc* should see what we do in the night." The entire room erupted with angry, disembodied voices. The members split into two camps, a vocal minority who condemned my presence, and the majority who claimed that the feast was a public event, a party open to anyone. Jean Baptiste lunged across the peristyle in our defense, and a semicircle of our supporters formed a cordon around us. My hands reached impulsively behind my back, tentatively gauging the strength of the thatch-and-mud wall. Beside me Rachel trembled.

"Don't worry," Josephine whispered, "there are no imbeciles here. It is a public ceremony. If it were private, you wouldn't be here. But this is only a party."

Suddenly the electric lights went off and on. Josephine looked concerned. Flames leaping from a fire at the base of the poteau mitan chased the shadows of the dancers across the walls of the temple, distorting them over the cracks in the thatch. A spasm of heavy breathing was broken by cries of intense effort, earth-pounding rhythms as the shoulders of the dancers moved like pistons, whiplash movements that pressed upon us.

"What is this stupidity?" Josephine yelled, even as she pulled us closer to her side. "Shanpwel! Open your asses!" The lights went out again, and the cordon around us tightened.

"Watch out!" Isnard screamed at us. "Breathe through your shirt!" For a single terrifying moment we waited, watching the dust suspended in the amber light. This was the time, Isnard believed, that the powder would come. I grabbed Rachel. All I could sense was past and future flowing over us like an uneddying river.

But nothing happened. Amid shrieks of excitement the lights came back on, and as the tension subsided Rachel's anxiety gave way to indignation.

"Listen," she said, "if they don't want us around here we should leave."

"No, my darling, you are not walking one step!" Josephine was adamant.

"But this is ridiculous!"

"Don't worry, child, it is past their time." And she was right. Whatever anger there had been but moments before and given way to laughter and public displays of welcome. The rotund president who had first objected to my presence now took the floor with a gesture of appeasement.

"Brothers and sisters," he began after the drums had stopped. "I'll explain something to you."

"Silence!" someone shouted.

"The question of a man being more a man or less a man, a man being white or a man being black, never changes whether he is right or wrong." He was interrupted by resounding applause.

"Myself," he went on, "I know that to want is to be able. I speak publicly now so that my motives will be understood. I took that cassette for a reason. President Jean Baptiste said it was his, and I gave it back to the *blanc*. But I am no fool. If the *blanc* comes to my place, he is welcome to record the vodoun part, and he may learn the songs of the Cannibal, but the words must remain ours. The songs are public, but the words are private. That is all. Excuse me, mademoiselle," he said looking toward Rachel, "I tell you this as a favor." Rachel thanked him and promised to take him up on his invitation to attend his ceremony, but her words were faint and lost in the growing pressure of impatient drummers and people anxious to dance.

All reflections and recriminations were dashed by the sudden intensity of motion that turned the peristyle into a great circle of spinning light. In a place that minutes before had been packed shoulder to shoulder there was now room for everyone to move. A dance of dervishes, of arms flailing the air, gave way to one of convulsive move-

ments, of feet pounding the ground and raising small clouds of dust
that lingered at knee level across the entire floor. Rachel is as good a
dancer as I am clumsy, but in the space of moments even I felt pulled
by the rhythm. To the delight of all, we leapt into line, moving as one
with the Shanpwel as they lunged across the ground. The staccato
beat of the drums flung us into the air, or beat us back into line. You
could actually feel its vibration striking at the base of your spine and
rising like electricity to your skull. For hours it seemed the drums
held us like that, sweat pouring from our skin, the smell of incense
and dust, of cheap perfume and rum fusing with a mocking sensuality
that brought our bodies so close before flinging them apart. My mind
wove its way through the night, becoming lost in some vast region of
the past, responding only to the steady rhythm, moving like a great
strand of kelp floating in a wild current yet all the time anchored
firmly to the earth.

I don't know when it ended. I only remember waking to the cool
wind outside the temple, and the dawn breaking through a pearl gray
sky. I remember the stillness of the palm trees, and on the road back a
few market women on their mules clip-clopping into town, and the
lights of Marcel Pierre's brothel, his clients leaning on the porch enjoying the aftertaste of a night in the darkness with those soft, pliant
women I had almost forgotten.

It wasn't until late the next afternoon that we were able to appreciate the full significance of the recording we had come so close to
losing. Amid the flurry of song and dance ritual, two vital pieces of
information stood out. First there was the speech. It was a clear public
statement that in explicit terms traced the origins of the Bizango directly to the Haitian revolution and the prerevolutionary leaders of
the Maroon bands, precisely in the way that the anthropologist Michel
Laguerre had proposed. Secondly, and perhaps more immediately important, the tape provided the names of more than a half-dozen prominent leaders in the Saint Marc region alone. With the generous assistance of Rachel's uncle Robert Erié, we were able in one day to establish
contact with five Bizango founding presidents, or emperors. One of
the most informative was also the most impressive, a crafty and powerful figure named Jean-Jacques Leophin.

Our first call on Leophin was purely social, and he received us
formally. He was an old man, thin, slightly stooped, with a penchant

for gold, which he displayed in wide bands on each finger and great loops of chains around his neck. A snap of his fingers brought forth a table and chairs, a tray of ice and whiskey. We drank in the middle of a blue-walled peristyle, surrounded by leaping images of the djab and the penetrating vision of the Black Virgin. In the yard outside, propped up on blocks like an icon, sat a broken-down Mercedes-Benz.

Beneath these and other frills—a black fedora and the habitual use of a cigarette holder—lay the soul, I would soon learn, of a man fully wed to the mystical. When he spoke his eyes sank deep into every listener, and from within his layered phrases came a resonance that lent his words great density. As we got to know him better he struck me as one of those rare people who have managed to forge a unique personal philosophy from a thousand disparate elements of the universe, and who then dedicate their lives to living up to the tenets of their belief. He spoke in parables, reciting myths and legends from Africa, mixing them liberally with citations from *Le Petit Albert*, a medieval text of sorcery once banned in Haiti, yet still the bible of choice of every conjurer and magician.

"Bizango is a word that comes from the Cannibal," he explained. "You find this word in the Red Dragon or the books of the Wizard Emmanuel. Bizango is to prove that change is possible. That's why we say 'learn to change.' We are in the world, and we can change in the world. Everyone says that the Shanpwel change people into pigs, but we say this only because it teaches that everything is relative. You may think that you and I are equals, are humans with the same skin, the same form, but another being looking at us from behind might say that we're two 'pigs,' or 'donkeys,' or even 'invisible.' This is what is called the Bizango changing. That is what it means.

"Long ago the Shanpwel did not exist. The four great Makala nations, that of the north, south, east, and west, formed the Bizango to create order and respect among their people."

"In the time before the *blanc?*"

"Yes, before and after. There was a girl from the south who kept taking the words of one leader and passing them around. She was the mouthpiece stirring up trouble. Finally, the four leaders came together, and once they realized what was going on they had to have her executed. A great public feast followed, and it was announced to the people that what happens within the four nations must stay there, and what happens outside stays outside. That's when they formed the

Bizango to protect against the dangers of the mouth. That's why even today if someone speaks against the society he must be made to respect his mouth. So we all come together."

"To form a judgment?"

"No, a council. Someone gets the feast organized—the rice, beans, colas, rum, and so on—and then the major members of the Bizango move apart. The groupe d'état-majeur—president, ministers, the queens. We pass the person. Once we agree that, say, so and so should die January 15, then we go to the emperor, in this region myself, and if the emperor agrees we mark the paper and the person dies. But if anyone disagrees, the case may drag on. Every three months or so we will convene a séance to discuss the case. If we reach an impasse we form a group of thirteen members, and the majority will rule. This is how the Shanpwel works. But a society cannot judge someone who hasn't directly affected a member. You can't just work against anyone."

"Is this what it means to sell someone to the society?"

"Only in that there is a justice. There is always a justice. But selling is different in that it is an action that starts from the people, not the leadership of the society. Selling is a means for each member of the society to seek his own justice. If you have a dispute, you make a small deposit of money and give your name and a description of the problem to the emperor, and if it is conscientious the society will pass both you and the offending party through a judgment."

"But does the body of the person appear before the society?"

"A stupid question. I tell you again and again, but you don't listen. I say it is a justice. Of course the accused appears. The emperor dispatches the chasseur, the hunter. That is his task. What kind of justice would judge a man from afar?" Leophin shifted in his chair impatiently. "But naturally these affairs may become complicated. Sometimes the one who has been sold may remain ill for as long as three years while the judgment is pending. In that case, since everyone knows that the sick one is on the list, the family may hire a diviner to go out and find out what happened. Once he does so, the family has one chance to make retribution. Word goes out and the society to whom the person was sold gathers. The family must pay each of the four leaders of the society."

"So the punished one may be bought back?"

"Under certain circumstances."

"Such as?"

"Look, I tell you it is a judgment, so the circumstances will vary."

"But what if I just want to get rid of someone?"

"You will be judged, for it will be you who breaks the code of the society."

"What code?"

"The seven actions." Without our asking, like a master of jurisprudence, Jean-Jacques Leophin enumerated the seven transgressions for which one could be sold to a society. These were, in order:

1. Ambition—excessive material advancement at the obvious expense of family and dependents.
2. Displaying lack of respect for one's fellows.
3. Denigrating the Bizango society.
4. Stealing another man's woman.
5. Spreading loose talk that slanders and affects the well-being of others.
6. Harming members of one's family.
7. Land issues—any action that unjustly keeps another from working the land.

The list read like a character profile of Clairvius Narcisse, and listening to it sent me whirling back to his case.

Clairvius fought often with various members of his own family. He sired numerous children whom he didn't support. By neglecting these and other community obligations, he managed to save enough money so that his house was the first in the lakou to have the thatch roof replaced by tin. But although his profligate existence certainly offended his extended family, the dispute with his brother was basically a question of access to land, and a dispute as serious as that between the Narcisse brothers would likely have involved arbitration by the Bizango society. Herard Simon had told me that it had been an uncle of Clairvius who had actually requested that the tribunal be convened. We can only surmise what might have occurred at that judgment, but there are several points to consider. The father of the two brothers was still alive at the time, and Clairvius Narcisse did not have any children recognized by the community. Now in Haiti it is customary that family land-holdings not be divided up "among the first generation of heirs, since younger brothers would not take it upon themselves to imply such disrespect of the senior ones as to demand that their tracts be split off." What's more, if heirs "insist on a formal division before the death of their elders, tradition brands them disrespectful and impertinent." One of the brothers was clearly wrong in terms of the proper code of conduct of the Haitian peasant society. The most important obligation of a patriarch in Haiti is to keep "family resources intact in order to provide a start in life for children." This explicit emphasis on posterity would have placed the childless, wifeless Narcisse in a less favorable position compared to the brother who had a large family to support at the time. Furthermore, if Narcisse had, in fact, been in the right and had been zombified by the guilty brother, it is difficult to imagine that the secret society would have permitted that brother to live on in peace in the village for close to twenty years. Given the current chilly relationship between Narcisse and his family, it seems more likely that Clairvius was the guilty party, a conviction held by the majority of my informants in Haiti and reinforced by an extraordinary statement made by Jean-Jacques Leophin the moment I mentioned Narcisse's name.

"Narcisse's brother sold him to a society of Caho. It was the seventh condition. It was the parent's land. He tried to take it by force." He paused, then added emphatically, "But this doesn't mean that the society is an evil thing. If someone on the inside of your house betrays you, they deserve death. But that doesn't make your house a bad place."

"So the people who took Narcisse were not from his village?"

"This happens. For example, I'm here at Fresineau. Everyone knows that Leophin is the master of the area. I have my limits from Gros Morne up to Montrouis. That's my quarter. Another society can't leave Archaie to catch someone in my territory unless he comes to me first. He'll explain the problem, and if it seems reasonable, I'll call a séance to discuss giving away the guilty one. If my people object, or if the accusation is unjust, I place him under my protection, and he won't be harmed. All the emperors communicate between each other, sometimes in person, sometimes by means of the superviseur, the messenger."

"The Bizango reaches into every corner of the land," Rachel said in honest amazement.

"It doesn't reach," Leophin corrected her. "*It is already there.* You see, we are stars. We work at night but we touch everything. If you are poor, I will call an assembly to cover your needs. If you are hungry, I will give you food. If you need work, the society will give you enough to start a trade. That is the Bizango. It is hand in hand.

"There is only one thing that the society refuses to get involved in. You can do almost anything, and the society may let you off, but if you stick your mouth in government talk, you can forget it. The society says you must respect the grade of the chief of the law. You can't just seize any office, you must deserve it. Go through elections and hear the voice of the people."

"But what if, for example, there is a policeman, abusing a member of the society," I asked.

"If you like the government, then you'd best remember that all the king's dogs are kings, even those doing evil. On the other hand, if the policeman is a real bother, one day his chief will call him in and tell him that he's been transferred. We don't want to harm him, but we too have our limits of tolerance."

"So the Bizango presidents work actively with the government. What happens when . . ."

Leophin interrupted somewhat indignantly. "The government cooperates with us. They have to. Imagine what would happen if some invaders landed in some remote corner of the Department of the Northwest. They would be dead before they left the beaches. But not by the hand of the government. It is the country itself that has been prepared for such things since ancient times.

"The people in the government in Port-au-Prince must cooperate with us. We were here before them, and if we didn't want them, they wouldn't be where they are. There are not many guns in the country, but those that there are, we have them."

This last statement of Jean-Jacques Leophin was no idle boast, at least if we are to judge by the zeal with which prominent national politicians, most notably Dr. François Duvalier, have courted the Bizango societies. The Duvalier revolution, often misrepresented in the Western press and remembered only for its later brutal excesses, began as a reaction on the part of the black majority to the excessive prominence of a small ruling elite that had dominated the nation politically and economically for most of its history. When Duvalier was first elected in 1957, he was unable to trust the army—indeed by the end of his tenure in office there would be over a dozen attempted invasions or coups d'état—and thus he created his own security force, the Volunteers for the National Security, or the Ton Ton Macoute, as they became known. (The latter name, incidentally, comes from a Haitian folktale that admonishes misbehaving children that their Ton

Ton, or uncle, will carry them off in his macoute, his shoulder bag.) To date, however, nobody has adequately explained the genesis of the Ton Ton Macoute as a national organization, or the remarkable speed with which it was established and emplaced in virtually every Haitian community. An explanation may lie in the network of the Bizango societies discussed so candidly by Jean-Jacques Leophin.

François Duvalier, a physician by training, was a keen student of Haitian culture. A published ethnologist, he was in his youth a pivotal member of a small group who put out an influential journal called *Les. Griots,* and it was within its pages that the germ of Duvalier's movement was sown. Though themselves scions of the elite, well educated and thoroughly urban, the intellectuals galvanized by *Les Griots* were responding to the humiliation of the American occupation and the flaccid acquiescence of their bourgeois peers. They did so by espousing a new nationalism that openly acknowledged the African roots of the Haitian people. At a time when drums and other religious cult objects were being hunted down and burned, and the peasants forced to swear loyalty to the Roman Catholic church, the members of *Les Griots* declared that vodoun was the legitimate religion of the people. It was a courageous stand, and one that earned Duvalier the unqualified support of the traditional society. During the 1957 election that brought him into power, Duvalier actively sought the endorsement of the houngan, and in certain sections of the country vodoun temples served as local campaign headquarters. With his success, François Duvalier became the first national leader in almost a hundred years to recognize the legitimacy of the vodoun religion and the rights of the people to practice it. During his term in office, he appointed houngan to prominent governmental positions. At least once he had all the vodoun priests in the country brought to the national palace to confer with him. And he was himself rumored to be a practicing houngan. It was an extraordinary transformation of official government policy. A year or two before the accession of Duvalier, vodoun drums were still being burned; a year after, a vodoun priest was serving as minister of education.

Critically, François Duvalier knew that he was surrounded by enemies, and he recognized with equal clarity that his strength and the ultimate power of any black president lay within the traditional society. Throughout his time in office, he went out of his way to penetrate the network of social control that already, as Jean-Jacques Leophin had suggested, existed within that society. He openly courted prominent houngan, and it is no coincidence that a man such as Herard Simon—a man both deeply religious and deeply patriotic—became the effective head of the Ton Ton Macoute for a full fifth of the country. Before Duvalier, blacks had limited access to public or governmental positions. In some cities there were parks where by unwritten agreement blacks were not permitted to walk. For men like Herard, Duvalier seemed like a savior. That was why, when I once asked him if he had had to kill many people during the early days of the struggle, he could reply sincerely, "I didn't kill any people, only enemies."

Undoubtedly Duvalier's close contacts with the houngan put him directly in touch with the Bizango societies and their leaders. Significantly, he was the first national president to take a direct personal interest in the appointment of each chef de section. It is also intriguing that the Haitian peasantry came to regard Duvalier as the personification of Baron Samedi, a spirit prominently associated with the secret societies. In his own dress and public behavior Duvalier appeared to affect that role quite deliberately: the ubiquitous black horn-rimmed glasses, the dark suit, and narrow black tie are the apparel that show up time and again on the old lithographs of the popular spirit. In short, whatever his motives, François Duvalier succeeded in penetrating the traditional vodoun society on a number of levels. The leaders of the secret societies almost inevitably became powerful members of the Ton Ton Macoute, and if the latter was not actually recruited from the Bizango, the membership of the two organizations overlapped to a significant degree. In the end, one might almost ask whether or not François Duvalier himself did not become the symbolic or effective head of the secret societies.

Josephine, the old woman who had befriended us that night at the ceremony, sold beans in the Saint Marc market, so she wasn't hard to find. By then we were recognized by most of the Shanpwel, and a dozen familiar but peculiarly anonymous faces guided us through the cluttered market stalls. It made for a strange sensation, weaving past the telling smiles, being known but still not knowing.

When we reached her shop—a simple wagon with an awning of tattered cloth, a rusty measuring tin, and a few neat piles of speckled beans—it was untended, but within moments Josephine came scampering back like a schoolgirl, so surprised and delighted that she could scarcely keep still. Leaving her business in the care of a neighbor, she marched us back through the market, pausing often to caper about the stalls of her friends until finally by the most circuitous route imaginable we reached the jeep. Then rather than taking the most direct road to the home of the president of her society—for that was the point of our visit—she managed to have us drive not once but twice through the market.

Eventually Josephine directed us along a dirt road that ran through the irrigated land east of Saint Marc. The valley, like so many along the central coast of Haiti, is an oasis in the midst of a barren man-made desert, and as we reached just past the edge, where the lush fields gave way to scrubland, we stopped. It was a melancholy place—a few tired trees leading up to a small compound literally carved into the porous sidehill. There were two main structures, both small, linked by a tonnelle and surrounded by a wattle fence. Though not old, the mud surface of the temple had already cracked into a thousand pieces. Overhead, a limp Haitian flag hung on a long staff.

Inside it was cooler, and as our eyes adjusted to the light, Josephine introduced us to her president, Andrés Celestin. As it happened we had seen him twice before, once at the ceremony where he had been presented with the other Bizango leaders, and a second time when Rachel's uncle Robert Erié had pointed him out in the main plaza of Saint Marc. Then he had appeared rather dashing, dressed in the stiff denim uniform of the Ton Ton Macoute. Now he seemed a broken man, lying prostrate on a cot with much of his face swollen and distorted by a sharp blow received, as we would learn later, when two Rara bands had met and clashed several nights previously. He was in no condition to receive anyone, and when despite his obvious pain he tried to stand to greet us, Rachel moved quickly to his side to ease him gently onto his back. He was severely concussed, and though the wound itself was not serious, it had become dangerously infected. We remained with him only long enough to promise to return the next day, and then, leaving some money for food, we hurried back to Saint Marc to purchase medication, which we dispatched with the old woman Josephine. We did return the next day, and each day after that until slowly his condition improved. Finally, about a week after being injured, he felt strong enough to speak with us, and by then, of course, he knew exactly what we needed.

Quite unlike Leophin or Jean Baptiste, established men whose authority was so certain that it appeared transparent, Andrés was a man on the rise, and his every gesture revealed a restless entrepreneurial spirit. As a youth he had deliberately moved close to Saint Marc, a center of Bizango activity, and it came as no surprise to discover later that he had been prominent in Leophin's society until eventually the competition between them became too great. Dismissed for disobeying orders, he had formed his own society, which he was now in the process of consolidating. He was ambitious, perhaps excessively so, but he wasn't corrupt, and in a grand manner he was terribly sincere. It was just that like many of his peers, while being true to his gods, he was more than willing to push them a bit toward satisfying his own aspirations. For Andrés, our chance meeting was a potent opportunity, for him no less than us.

"It is quite normal," he explained, "that we work together. I have something that you want, which is knowledge. And you have something that I need." He was speaking of more than just money, but also what our connections could mean for his society, and for my part I found his summation refreshingly frank. Behind him a pod of children gathered around a cooking fire, pushing out their hungry bellies. To one side, a tired-looking woman dislodged a clump of coarse sand from the ground to scour a pot; beside her lay piles of spindly firewood, and tin cans of water too precious to bathe in. A certain undeniable truth lay between us. We eyed each other for a moment, and then he lifted his head from the cot and his croaky laughter sealed our agreement.

It was Wednesday, the day when the society met, and that night following Andrés's instructions we returned to the compound. We were late, as usual, and already most of the members were there, mingling about, waiting somewhat impatiently in the dark for a Coleman lantern so that the ceremony could begin. At the back of the tonnelle, in place of the cooking fires, three women moved impulsively to the rhythm of a small battery of drums. They stopped soon after we arrived. The compound was too small and the members were too few for anyone to feign discretion. Those who knew greeted us fondly; the rest stared in bemused disbelief.

Andrés was completely unfazed by our awkward arrival, and after a few moments of idle conversation he suggested politely that it was a good time to speak with the master. Groping in the darkness, we followed him out of the shelter of the tonnelle and around the corner of his house to a small outbuilding perched precariously on the steepest part of the sidehill. After knocking three times to alert the spirits, Andrés unlocked the rusty latch and led the way inside, passing through a voile curtain that concealed a small altar in the midst of which burned a single wick in a bowl of hot wax. Above the flame, a canopy of dozens of small mirrors and bells hung by ribbons from the ceiling. Rachel and I sat close together, knees touching on two small wicker chairs huddled to one side. Andrés leaned toward us, and for the first time I noticed the black patch that covered his wounded eye.

"You see, you have never met him, and you must. I cannot do something that is beyond my time. He is the one who can really give you the secrets." Someone coughed behind us, and I realized we were not alone. Andrés pulled back the curtain and ordered whoever it was to bring a bottle of rum. "And tell the people they can dance," he added before turning back to the altar. "Now, my friends, there are so many lessons. I shall show you great things. Wade, you are a *blanc*, and Rachel, you must serve as his master to show him all that I show you. Understand? Both of you? Good."

Andrés picked up a bell and rang it while his free hand swept through the mirrors, sending bits of amber light dancing across every surface. The assistant scuttled back in, placed the bottle at our feet, and backed away. Another chime, and then Andrés sat down, and commenced a hypnotic drone of the liturgy, on and on, beseeching perhaps two hundred or more spirits and powers, until, stumbling over a single syllable, his voice changed. And started to stutter almost uncontrollably. In the guise of the master "Hector Victor," the *pwin*, or mystical force, of the society had arrived.

"O-O-O-O-O Wha-wha-what is this? How are you? We-we-we-we have two foreigners here. Where are you from? Port-au-Prince? I-I-I'm glad. I am Hector Victor. I serve anywhere, do-do anything. Listen, little lady. I-I-It seems you need me? You need information? No? What then? O-O-Oh! Ha! So the guy talked to you already. What did he tell you?"

"He said that we should speak with you because you're the one who knows," Rachel answered quietly.

"O-O-Oh! It is true. I do know one thing. That is something that costs. It's like school. Y-You must pay. Y-Y-You see. A-Ahow much can you pay?"

"No, no, Hector Victor, you must tell us the price. I can't just say any amount."

"Oh." The voice paused. "Sa-sa-say, is that rum?" I passed him the bottle. Only in Haiti, I realized, is it possible to drink rum and haggle with a god. "We-we-well," Hector Victor continued, "little lady, myself, I know that you can make me much more money than what I make here. So, th-th-this will cost s-s-s-sixty dollars. But you'll see how worthy it is. Your father is a houngan, no? Y-y-you'll see how pleased he'll be."

"Well," Rachel paused as if buying vegetables.

"Rachel," I said quietly, "it's fine."

"All right, we'll pay forty dollars now and owe the rest."

"Oh. Yes, I can trust you. I surely can. I-I-I'll take this, but you'll see that it is worth far more. So, I'll be going now and let my horse do the work. But remember that Hector Victor can mount any bagi." Hector Victor turned to the assistant that had brought the rum and handed him one of the bills I had given him. "T-t-t-take this and change it and give it to the others. Tell them they'll bring more." The shadowy figure of the assistant scurried out of the bagi.

The master blew his nose into a piece of newsprint, and then sat slumped on his chair looking comically disconsolate. "M-m-m-money," he sighed, "see how it goes?" Another assistant arrived. "W-w-w-who's there? Ah! My dear. Listen, there is something we are going to do. Do it well. We need these people. W-w-w-we want them and nothing must go wrong. T-t-t-tell my horse that these people will make us walk over land we never could have walked over." The assistant distracted us for a moment, insisting that we write down his name so that we could contact him by way of the public announcements on the national radio. The master lifted his one good eye into the light, and once he was sure that the correct information had been recorded, he disappeared, leaving a limp body that soon came to life in the form of Andrés Celestin.

Andrés was exhausted, and it took a few minutes and several drinks of rum before he could pull himself together. Outside the bagi, the rhythm of the drums had changed and the hollow sound of the conch shell signaled the beginning of the séance.

"So," the president said finally, "now we can get to work. Did Hector Victor speak properly? Good. He can be very rude. Do you have any matches? Stand up, both of you." Andrés blew out the candle, and the tip of the wick smoldered like a jewel in the darkness before it died.

"This is a moment that is still to come. A hard time when you'll get very dirty. The clothes you arrive in will be worth nothing when you leave. It is rough, perhaps too rough." He paused, and I could hear his breathing as he moved away from the altar. "Light the flame!" A match flared and moved toward the candle.

"The candles are stars. Everything falls into flux and day becomes night, the end becomes the beginning. This is the new life. Move away from the light."

Rachel and I stepped cautiously into the shadows behind the curtain. Andrés took each of us, first Rachel, then me, and taught us how to shake hands, and how to greet the members of the Bizango. Once satisfied, he led us out of the bagi back to the tonnelle. By then the gate of the compound was sealed and guarded, and the Shanpwel had changed skins and fallen into a single rank that ran along the periphery of the enclosure. Andrés called them to attention, then began to speak. Almost every phrase earned him immediate and violent applause. He explained that he had met us at the celebration of Empress Adèle, and that it had infuriated him to have seen us treated so poorly. How many fools could there be, he asked, who did not know that the whites could be important, that they could help when the proper time came?

"The whites need us, and we need them." Those were his final words, and with them he instructed Rachel and me to shake hands with each member. Slowly, obeying his order, we moved around the ring offering the ritual greeting we had just learned. Some of the members smiled, some giggled, some displayed faces as hard as wood.

The drums began, then stopped, and one of the queens, her voice like a peacock, began the sad plaintive hymn of the adoration. Other women opened the wooden doors of a small earthen chamber and brought out the sacred coffin, the *madoulè*, the mother of the society. The procession formed, and Andrés added us to its tail, and together with the other members of the Shanpwel we moved in an ever narrowing circle, so slowly that my ankles began to tremble. When the coffin finally came to rest, each member stepped before it, made his or her offering, and saluted. A series of genuflections followed, and there were other lessons. And then, once the coffin had been lifted and marched back to its chamber, Andrés again led us away from the tonnelle to the bagi.

"It is only a beginning," he told us, "like a shadow of what there is to know." We would return, he insisted, both to host a feast and to discover other mysteries. After that it would be our duty to come to each séance. Even if I, in particular, was displaced beyond the great water, I would have to send my genie as an emissary—also, of course, the occasional financial offering to hold up those who might be in need. This said, Andrés leaned forward and whispered two passwords that we would have to know. Then, reaching behind him, he brought out four glasses.

"Drink," he said, handing one to each of us. The green potion seared my throat.

"And this." The second dose was viscid and syrupy. "Now you will know how things are to be," he explained. "The first it was bitter. And the second?"

"Sweet," I replied.

"Yes. Well, this is the Shanpwel. One side is bitter, the other is sweet. When you're on the bitter side, you won't know your own mothers." The president, his silhouette showing through the voile curtain, threw the dregs of each glass onto the ground, and the drops ran together like mercury to sink into the dusty surface of the earth floor.

After that, Rachel and I entered a long, silent passage and for the next month were as strangers to the sunlight.

Epilogue

"So is this your choice?" Herard asked. We had met on a roadbed beneath a searing sun. I was ill at the time, my body frail and my throat too dry for talk, and the fevers pushed me into the shade of the scrubland. Herard followed.

"You didn't listen to me," he said. "You cannot do these things casually. Everything carries its price. There are things to know and things best left unknown."

"But we have been received well," I told him.

"Yes, a child's game. What do you know?"

"I know that the answers I seek will be found among the Bizango."

"You know nothing." He looked away. "And when you fall, will the society of this man be strong enough to lift you up?" The words startled me, for they were the very same as those used by Jean-Jacques Leophin; you could join but one Bizango society, he had cautioned, and it had best be one that could support you in time of need.

"You have been lucky," Herard continued. "Do not wait for a time when you'll see things that you should not see."

"Do you mean the sacrifices?"

"That is the time. Near the end of each year they must gather their powers."

Herard turned and began to make his way slowly back to the road. As he reached the jeep he called back to me, "You may choose to join the Bizango, or you may come with me and serve the loa. You cannot do both."

Here were two diverging paths that represented less a choice than a signal that my work had come to a temporary end. To the last Herard would condemn the Bizango, and what he had told me contained a deep warning, a powerful declaration that my actions bore consequences. To complete the initiation into the Bizango society of Andrés Celestin would be an irrevocable step. No longer would I be an outsider, free to alight wherever I chose. I would become part of a matrix, bound to the other members by vows and obligations.

I had arrived in Haiti to investigate zombis. A poison had been found and identified, and a substance had been indicated that was chemically capable of maintaining a person so poisoned in a zombi state. Yet as a Western scientist seeking a folk preparation I had found myself swept into a complex worldview utterly different from my own and one that left me demonstrating less the chemical basis of a popular belief than the psychological and cultural foundations of a chemical event. Perhaps more significantly, the research had suggested that there was a logical purpose to zombification that was consistent with the heritage of the people.

To be sure, I had failed to document a zombi as it was taken from the cemetery, but this was no longer something I deliberately sought out. In the last weeks I had, in fact, been offered two promising opportunities to do so, provided, of course, that the cash payment to the bokor be sufficient. I had gone as far as making the preliminary contacts before I realized that the whole concept had changed. A year or two before I would have gone ahead, emboldened by the deep skepticism I had brought with me to Haiti; now that I had completely overcome my doubts, I found myself forced by that very certainty to turn aside an opportunity that might have offered final proof to those who still shared my early skepticism. The money that the party in question demanded was considerable. If the affair turned out to be fraudulent, I would have wasted the money. If, on the other hand, it turned out to be legitimate, I would have no way of being certain that the money had not been responsible for the victim's fate. It was an ethical Rubicon I was not willing to cross.

What did remain were the secret societies, but to pursue that connection demanded a new orientation. It would not be enough to document a set of principles as perceived by the Bizango leadership; I would have to observe how they were played out in the day-to-day lives of the people. To do that I would have to enter a community and undertake a study that would require months, not weeks. It was another level of commitment that might well involve completing my initiation into the Bizango society—not as an end in itself but only as a means. It was not something to be undertaken lightly. Regretfully, I decided to curtail my work with Andrés.

It was, in any case, a time for change. Rachel had reports to prepare before she returned to her university and had to stay in the capital. I had a book to write. But it was summer again, the season of the pilgrimages, and just before returning to the States I was drawn north to the festival at Plaine du Nord, where those who serve the loa pay homage to Ogoun, the god of fire and war.

A sacred mapou tree marks the place where once each year a mud pond spreads over a dry roadbed near the center of the village. Like the waters of Saut d'Eau, the mud of the basin is said to be profoundly curative, and each year thousands of pilgrims arrive, some to fill their bottles, some to cleanse their babies, many to bathe. Unlike Saut d'Eau, the area around the basin is constricted by houses that funnel all the energy of the pilgrims into a small, intensely charged space. And in place of the serenity of Damballah, there is the raging energy of Ogoun.

Around the periphery of the basin a ring of candles burns for the spirit, and the pilgrims dressed in bright cotton lean precariously over the mud to leave offerings of rum and meat, rice and wine. There is a battery of drums to one side, and those mounted by the spirit enter the basin, disappear, and emerge transformed. One sees a young man, his body submerged with only his eyes showing, move steadily like a reptile past the legs of naked women, their skin coated with slimy clay. Beside them, children dive like ducks for tossed coins. At the base of the mapou, Ogoun feeds leaves and rum to a sacrificial bull; others reach out to touch it, and then the machete cuts into its throat, and the blood spreads over the surface of the mud.

I was watching all this when I felt something fluid—not water or sweat or rum—trickle down my arm. I turned to a man pressed close beside me and saw his arm, riddled with needles and small blades, and the blood running copiously over the scars of past years, staining some leaves bound to his elbow before dripping from his skin to mine. The man was smiling. He too was possessed, like the youth straddling the dying bull, or the dancers and the women wallowing in the mud.

Glossary

ADORATION   The song that accompanies the offering of money at ceremonies.

AGWÉ   Vodoun loa; the spirit of the sea.

APHONIA   Loss of voice due to functional or organic disorder.

ASSON   The sacred rattle of the houngan or mambo; a calabash filled with seeds or snake vertebrae and covered by a loose web of beads and snake vertebrae.

AUTONOMIC NERVOUS SYSTEM   That part of the nervous system regulating involuntary action, as of the heart and digestive system.

AYIDA WEDO   Vodoun loa whose image is the rainbow; mate of Damballah.

AYIZAN   Vodoun loa, patroness of the marketplace and mate of Loco.

BAGI   Inner sanctuary of the temple, the room containing the altar for the spirits.

BAKA   An evil spirit, a supernatural agent of the sorcerers that often takes the form of an animal.

BARON SAMEDI   Vodoun loa, the lord and guardian of the cemetery, represented there by a large cross placed over the grave of the first man to be buried there. Important spirit of the Bizango.

BÊTE SEREINE   Animal of the Night, an appellation for the members of the Bizango societies. Literally, "serene beast."

BIZANGO   The name of the secret society; also connotes the rite practiced by the Shanpwel. The name is possibly derived from the Bissagos, an African tribe that occupied an archipelego of the same name off the coast of Kakonda, between Sierra Leone and Cape Verde.

BLANC   "White," but more precisely "foreigner," and like the term *gringo* in parts of Latin America, not necessarily pejorative.

BOKOR   Although the word is derived from the Fon word *bokono*, or priest, in Haiti the bokor has come to represent the one who practices sorcery and black magic as a profession.

BOUGA   The toad *Bufo marinus*.

BOURREAU   The executioner, a rank in the Bizango societies said to be responsible for enforcing the decisions of the groupe d'état-majeur.

CACIQUE   Term is still used in Haiti to refer to the pre-Columbian Amerindian rulers and divisions of the country.

CALABASH   A gourd used as a water container or ceremonial vessel.

CAMIONNETTE   A small truck used as a bus.

CANARI   A clay jar that is used for sheltering the ti bon ange, and which is broken at funeral rites.

CANNIBAL   A term for a secret society.

CANZO   The ordeal by fire through which the adept passes into initiation.

CARREFOUR   The crossroads; also a vodoun loa and one associated with the Bizango, as well as the Petro rites.

CATA   The smallest of the set of three Rada drums.

CHASSEUR   The hunter, a rank of the Bizango society.

CHEF DE SECTION   The appointed authority responsible for policing the *section rurale*.

CHEVAL   The horse; in vodoun parlance the individual who is mounted by the spirit. Hence the metaphor for possession and the meaning of the title of Maya Deren's book *The Divine Horsemen*.

CHRISTOPHE, HENRI   The former slave, lieutenant of Dessalines, who ruled the northern half of Haiti from 1807 to 1820, when he took his own life.

CLAIRIN   Inexpensive clear white rum used commonly in vodoun ritual.

COMBITE   Collective labor party, generally for agricultural work.

CONCOMBRE ZOMBI   The vernacular name for *Datura stramonium*.

CONVOI   The name of a particular secret society.

CORPS CADAVRE   The body, the flesh, and the blood as opposed to the various components of the vodoun soul.

COUP L'AIRE   An air spell, a means of passing a magical spell that will cause misfortune and illness.

COUP N'ÂME   A soul spell, a means of magically capturing the ti bon ange of an individual.

COUP POUDRE   A powder spell, a magical powder that may cause illness and/or death.

CREOLE   The language of the traditional vodoun society, also used to designate anything native to Haiti.

CYANOSIS   A bluish coloration to the skin caused by lack of oxygen in the blood.

DAMBALLAH WEDO   Vodoun loa whose image is the serpent, the mate of Ayida Wedo.

DATURA   A genus of plants in the potato family (Solanaceae).

DESSALINES, JEAN-JACQUES   Leading general of Toussaint L'Ouverture and first president of independent Haiti from 1804 until his assassination in 1806.

DESSOUNIN   The ritual separating a dead person's ti bon ange and spirit, or loa, from the body.

DJAB   The devil, a baka, a malevolent force.

DOKTE FEUILLES   The leaf doctor, herbalist healer.

DYSPHAGIA   Difficulty in swallowing.

EMPEREUR   A founding president of a Bizango society; also the titular leader of a number of different societies.

ERZULIE   A vodoun loa, the spirit of love.

ESPRIT   The spirits or soul of the dead.

FUGU   A genus of marine fish and the vernacular name for the tetrodo-toxin-containing fish that are served in Japanese restaurants.

FWET KASH   A sisal whip.

GAD   A protective charm, a tattoo that is physically applied to the skin at initiation to protect the individual from evil.

GOURDE   Unit of Haitian money worth twenty cents.

GOVI   The sacred clay vessels in which the spirits of the dead or the loa are housed.

GRANS BWA   A vodoun loa, the spirit of the forest.

GROS BON ANGE   "Big good angel"; that aspect of the vodoun soul shared by all sentient beings; one individual's part of the vast pool of cosmic energy.

GROUPE D'ÉTAT-MAJEUR   Said to be executive leadership of the Bizango society.

GUEDE   A vodoun loa, the spirit of the dead.

GUINÉE   Africa, or the mythical homeland; the land of the loa.

HOODOO   A variation of the word *voodoo* used commonly in the southern United States.

HOUNFOUR   The vodoun temple, including the material structure and the acolytes that serve there. When used in opposition to peristyle, it refers to the inner sanctuary with the altar.

HOUNGAN   Vodoun priest.

HOUNSIS   Members of the société, or hounfour, at various levels of initia-tion. From the Fon language (*hu*—a divinity; *si*—a spouse).

INVISIBLES, LES   Term for all the invisible spirits including the loa.

LAMBI   Conch shell used as a trumpet.

LANGAGE   Sacred language used only in ceremonies of African origin.

LEGBA   A vodoun loa, the spirit of communication and the crossroad.

LEOPARD SOCIETY   The secret society of the Efik of Old Calabar.

LOA   The deities of the vodoun faith.

LOCO   A vodoun loa, the spirit of vegetation.

Loup garou   The werewolf. The flying queen, or reine voltige, of the Bizango is said to be a loup garou.

Macandal, François   Mandingue slave born in Africa and believed to have been executed in 1758 for his part in a poison conspiracy; also in contemporary Haiti the name of a secret society.

Macoute   Straw shoulder bag of the Haitian peasant.

Madoulè   The sacred coffin and symbol of the Bizango societies.

Maît'   Master, as in the dominant loa.

Malfacteur   Evildoer, particularly an individual who specializes in powders.

Maman   The mother, the largest of the three drums in the Rada battery.

Mambo   Female vodoun priestess.

Mangé Moun   "To eat people," a euphemism for killing someone.

Mapou   The sacred tree of the vodoun religion, *Ceiba pentandra* of the botanical family Bombacaceae.

Maroon   Fugitive slave, from Spanish root *cimarrón*, meaning "wild, unruly."

Mort bon Dieu   A call from God, a natural death.

Mystères   The loa.

Nam   Generic term derived from the French *âme*, or soul, and referring to the complete vodoun soul including gros bon ange, ti bon ange, and the other spiritual components.

N'âme   The spirit of the flesh that allows each cell to function.

Nèg Guinée   A person of Africa, of the mythical homeland, of African descent.

Ogoun   A vodoun loa, the spirit of fire, war, and the metallurgical elements; the blacksmith god.

Paquets Congo   A small sacred bundle containing magical ingredients that serves to protect a person against illness or evil; the closest object there is to the notorious and often misrepresented "voodoo doll."

Paresthesia   An abnormal sensation, as of prickling of the skin.

Peristyle   The roofed, usually unwalled area where most ceremonies occur.

Petro   A group of vodoun loa traditionally said to be of American origin, now increasingly believed to be derived from Congolese rites.

Pierre tonnerre   The thunderstones said to be created by the spirits and thus imbued with mystical healing powers.

Poro   A secret society among the Mende of Sierra Leone.

Poteau mitan   The centerpost of the peristyle and the axis along which the loa rise to enter the ceremonies.

Poudre (poud)   Magical powder.

President   The highest position below that of the emperor within the Bizango societies. Member of the groupe d'état-majeur.

Psychoactive   Designates a drug that has a specific effect on the mind.

Psychopharmacology   The study of the actions of drugs on the mind.

Pulmonary edema   An abnormal accumulation of fluid in the lungs.

PWIN   Magical power or force summoned to execute the will of the sorcerer or Bizango society; may be used for good or evil.

RADA   Vodoun rite; a body of loa, songs, and dances of Dahomean origin. Name derived from the town of Arada in Dahomey, now Benin.

RARA   Festival during the spring characterized by processions associated with particular hounfour or Bizango societies.

REINE   The queen: high-ranking female position in the Bizango societies. Première reine, the first queen, is a member of the groupe d'état-majeur. Below her are the deuxième reine (second queen), troisième reine (third queen), reine drapeau (flag queen).

REINE VOLTIGE   The flying queen; also conceived of as a loup garou, or werewolf. The four reines voltiges carry the sacred coffin during the formal Bizango processions.

SAINT DOMINGUE   The French colony that later became Haiti.

SÉANCE   The term used to describe the nocturnal gatherings of the Bizango.

SECONDE   The second or middle drum in the Rada battery of three.

SECRÉTAIRE   The secretary; rank of the Bizango society.

SECTION RURALE   The fundamental administrative unit of local government in rural Haiti.

SENTINELLE   The sentinel; rank of the Bizango society; the guard or scout who moves ahead of the processions and secures the entryway into the Bizango ceremonies.

SERVI LOA   The term used by vodoun acolytes to refer to their faith, "to serve the loa."

SERVITEUR   One who serves the loa.

SHANPWEL   Term used to refer to the secret societies; sometimes used interchangeably with Bizango, but more properly refers to the members of the Bizango, not the rite itself.

SOBO   Vodoun loa, the spirit of thunder.

SOCIÉTÉ   A hounfour and its members, not to be confused with Bizango society.

SOLDAT   The lowest rank in the Bizango society.

SUPERVISEUR   Rank of the Bizango society; said to be responsible for conveying messages between different society leaders in different regions of the country.

TAP TAP   Popular Creole term for the kaleidoscopic buses.

TETRODOTOXIN   Potent neurotoxin found in puffer fish and various other animals that blocks the conduction of nerve signals by completely stopping the movement of sodium ions into cells.

TI BON ANGE   That aspect of the vodoun soul said to be responsible for creating a person's character, willpower, and individuality.

TI GUINÉE   A child of Guinée, the mythical homeland; also used to refer to a member or offspring of the vodoun society.

TONNELLE   A thatch or corrugated tin roof improvised in the absence of a complete peristyle; a canopy beneath which ceremonies and dances occur.

TON TON MACOUTE   From ton ton, or uncle, and macoute, the straw shoulder bag of the peasant. Name for the independent security forces established by Dr. François Duvalier.

TOPICALLY ACTIVE   Designates a chemical or pharmaceutical substance that may be applied effectively to the skin.

TOUSSAINT L'OUVERTURE   Ex-slave, revolutionary leader, general, liberator historically perceived as the Simon Bolívar of Haiti. Betrayed by Napoleon and deported to France, where he died in 1803.

UREMIA   Toxic condition caused by the presence in the blood of waste products normally eliminated in the urine.

VÉVÉ   Symbolic designs drawn on the ground with flour or ashes and intended to invoke the loa. Each spirit has a characteristic vévé.

VLINBLINDINGUE   A name of a secret society (*also* vinbrindingue).

VODOUN   The theological principles and the practice of the Haitian traditional society.

VSN   Acronym for the Volontaires pour la Sécurité Nationale, the militia established by François Duvalier to protect his regime.

WANGA   A magical charm used for selfish or malevolent intent.

WÉTÉ MO NAN DLO   "To take the dead from the water"; the ritual whereby the ti bon ange is reclaimed by the living and given a new form.

WHITE DARKNESS   The term Maya Deren used to describe the state of possession that she experienced.

ZOMBI ASTRAL   A zombi of the ti bon ange. An aspect of the soul that may be transmuted at the will of the one who possesses it.

ZOMBI CADAVRE   The corps cadavre, gros bon ange, and the other spiritual components. A zombi of the flesh that can be made to work.

ZOMBI SAVANE   An ex-zombi, one who has been through the earth, become a zombi, and then returned to the state of the living.

Annotated Bibliography

1. THE JAGUAR

An account of the Darien expedition appears in Sebastian Snow's *The Ruck-Sack Man*, Hodder and Stoughton, London, 1976. For biographical background on Professor Schultes, see Krieg, M. B., *Green Medicine*, Bantam Books, New York, 1964. Schultes's most important work, written with Albert Hofmann, is *The Botany and Chemistry of Hallucinogens*, Charles C Thomas, Springfield, IL, 1980. Their popular treatment of the subject is *Plants of the Gods*, McGraw-Hill, New York, 1979.

2. "THE FRONTIER OF DEATH"

The difficulty of diagnosing death is discussed in: Kastenbaum, R., and R. Aisenberg, *The Psychology of Death*, Springer Publishing, New York, 1972; Mant, A. K., in Toynbee, A., ed., *Man's Concern with Death*, Hodder and Stoughton, London, 1968; Watson, L., *The Romeo Error*, Hodder and Stoughton, London, 1974. The case from the Sheffield mortuary was reported in *The Times*, London, 28 February 1970. The New York case is cited in Watson (1974), p. 15. Medical studies describing the ability of Indian fakirs include: Anand, B. K., G. S. Chhina, and B. Singh, Studies on Shri Ramanand Yogi during his stay in an air-tight box, *The Indian Journal of Medical Research* 49, no. 1 (1961): 82–89; Anand, B. K., and G. S. Chhina, Investigations on Yogis claiming to stop their heart beats, *The Indian Journal of Medical Research* 49, no. 1 (1961): 90–94. The case of Clairvius Narcisse was first introduced in Douyon, L., Les zombis dans le contexte vodou et Haitien, *Haiti Santé* 1 (1980): 19–23. The various issues associated with anesthesiology are presented in Orkin, F. K., and L. H. Cooperman, eds., *Complications in Anesthesiology*, J. B. Lippincott Co., Philadelphia, 1982.

3. THE CALABAR HYPOTHESIS

The San Pedro healing ceremony is described in Sharon, D., The San Pedro cactus in Peruvian folk healing, in Furst, P. T., ed., *Flesh of the Gods: The Ritual Use of Hallucinogens*, 114–35, Praeger, New York, 1972; and Sharon, D., *Wizard of the Four Winds*, The Free Press, New York, 1978. My own observations appear in Davis, E. W., Sacred plants of the San Pedro cult, *Botanical Museum Leaflets—Harvard University* 29, no. 4 (1983): 367–86.

The catalog referred to was Moscoso, R. M., *Catalogus Florae Domingensis* Pt. 1, New York, L & S Printing, Inc., 1943. Other information on *concombre zombi* was found in Brutus, T. C., and A. V. Pierre-Noel, *Les plantes et les légumes d'Haiti qui guérissent*, 3 vols., Imprimerie de l'Etat, Port-au-Prince, 1960.

For information on other aspects of datura see: Schleiffer, H., *Sacred Narcotics of the New World Indians*, Hafner Press, New York, 1973; Schleiffer, H., *Narcotic Plants of the Old World*, Lubrecht & Cramer, Monticello, NY, 1979; Hansen, H. A., *The Witch's Garden*, Unity Press, Santa Cruz, 1978; Weil, A. T., *The Marriage of the Sun and the Moon*, Houghton Mifflin Co., Boston, 1980; Emboden, W., *Narcotic Plants*, Macmillan Publishing Co., New York, 1979; Lockwood, T. E., The ethnobotany of *Brugmansia*, *Journal of Ethnopharmacology* 1 (1979): 147–64.

An excellent article on the Calabar bean is: Holmstedt, B., The ordeal bean of Old Calabar: the pageant of *Physostigma venosum* in medicine, in Swain, T., ed., *Plants in the Development of Modern Medicine*, Harvard University Press, Cambridge, 1972. For the secret societies of the Efik see: Forde, D., *The Efik Traders of Old Calabar*, Oxford University Press, London, 1956.

4. WHITE DARKNESS AND THE LIVING DEAD

Basic references on the vodoun religion include: Métraux, A., *Voodoo in Haiti*, Schocken Books, New York, 1972; Deren, M., *The Divine Horse-*

men—the Living Gods of Haiti, Thames and Hudson, London, 1953; Herskovits, M. J., *Life in a Haitian Valley,* Alfred A. Knopf, New York, 1937. See also the academic papers of George Simpson, in particular: Simpson, G. E., *Religious Cults of the Caribbean: Trinidad, Jamaica and Haiti,* Caribbean Monograph Series 15, Institute of Caribbean Studies, University of Puerto Rico, Rio Piedras, P.R., 1980; Marcelin, M., *Mythologie Vodou,* Les Editions Haïtiennes, Port-au-Prince, 1949; Maximilien, L., *Le Vodou Haitien,* Imprimerie de l'Etat, Port-au-Prince, 1945; Rigaud, O. M., The feasting of the gods in Haitian vodu, *Primitive Man* 19, nos. 1–2 (1946): 1–58. Another basic source is Courlander, H., *The Drum and the Hoe: Life and Lore of the Haitian People,* University of California Press, Berkeley, 1960. For vodoun song see: Courlander, H., *Haiti Singing,* Cooper Square Publishing, New York, 1973; and Laguerre, M. S., *Voodoo Heritage,* Sage Library of Social Research 98, Sage Publications, Beverly Hills, 1980. Two enjoyable and informative books based on personal accounts are: Dunham, K., *Island Possessed,* Doubleday & Co., Garden City, NY, 1969; and Huxley, F., *The Invisibles—Voodoo Gods in Haiti,* McGraw-Hill, New York, 1966. Of historical interest is Price-Mars, J., *Thus Spoke the Uncle,* Three Continents Press, Washington, D.C., 1983.

A case of another zombi from Ennery, the village where Ti Femme was discovered, is reported in Simpson, G. E., Magical practices in northern Haiti, *Journal of American Folklore* 67, no. 266 (1954): 401. Reports from Seabrook, W. B., *The Magic Island,* George G. Harrap & Co., London, 1929; and Hurston, Z., *Tell My Horse,* Turtle Island, Berkeley, 1981, are reviewed in Métraux, A. (1972). The cases of Ti Femme and Narcisse have been summarized in: Diederich, B., On the nature of zombi existence, *Caribbean Review* 12, no. 3 (1983): 14–17, 43–46. A regrettably sensational account of Narcisse appears in: Pradel, J., and J. Casgha, *Haiti: La République des morts vivants.* Editions du Rocher, Paris, 1983.

For accounts of firewalking see: Weil, A. T., *Health and Healing—Understanding Conventional and Alternative Medicine,* Houghton Mifflin Co., Boston, 1983.

5. A LESSON IN HISTORY

Basic historical sources include: Leyburn, J. G., *The Haitian People,* Yale University Press, New Haven, 1941; James, C. L. R., *The Black Jacobins,* Random House, New York, 1963; Moreau de Saint-Mery, *Description topographique, physique, civile, politique, et historique de la partie française de l'Isle de Saint Domingue,* Librairie Larose, Paris, 1958. Other early sources include: Baskett, J., *History of the Island of St. Domingo from Its First Discovery by Columbus to the Present Period,* Negro Universities Press, Westport, CT, 1971 [1818]; Brown, J., *The History and Present Condition of St. Domingo,* 2 vols., Frank Cass, London, 1971 [1837]; Franklin, J., *The Present State of Hayti,* Negro Universities Press, Westport, CT,

1970 [1828]; MacKenzie, C., *Notes on Haiti,* 2 vols., Frank Cass, London, 1971 [1830]; Wimpffen, F., *A Voyage to Saint Domingo,* London, 1817.

The contemporary social structure of Haiti is examined in: Horowitz, M. M., *Peoples and Cultures of the Caribbean,* The Natural History Press, 1971; Mintz, S., ed., *Working Papers in Haitian Society and Culture,* Antilles Research Program, Yale University, New Haven, 1975; Moral, P., *Le Paysan Haitien,* G. P. Maisonneuve et Larose, Paris, 1961; Simpson, G. E., Haiti's social structure, *American Sociological Review* 6, no. 5 (1941): 640–49; Laguerre, M., The place of voodoo in the social structure of Haiti, *Caribbean Quarterly* 19, no. 3 (1973): 36–50.

Perhaps the most illuminating analysis of the evolution of the contemporary peasant society is the work of Murray, G. F., "The Evolution of Haitian Peasant Land Tenure: A Case Study in Agrarian Adaptation to Population Growth," Ph.D. dissertation, Columbia University, 1977. Also of particular importance to me was: Fouchard, J., *The Haitian Maroons— Liberty or Death,* Edward W. Blyden Press, New York, 1981.

6. EVERYTHING IS POISON, NOTHING IS POISON

A good introduction to African religion may be found in Mbiti, J., *African Religions and Philosophy,* Anchor, New York, 1970. See also Parrinder, G., *West African Religion,* The Epworth Press, London, 1961. For a wonderful semipopular book see Watson, L., *Lightning Bird—The Story of One Man's Journey into Africa's Past,* Simon and Schuster, New York, 1982.

7. COLUMNS ON A BLACKBOARD

Notes on the plant ingredients are mentioned in a variety of sources including: Dalziel, J. M., *Useful Plants of West Africa,* London, 1937; Githens, T. S., *Drug Plants of Africa,* African Handbook 8, University of Pennsylvania Press, 1948. Sofowora, A., *Medicinal Plants and Traditional Medicine in Africa,* John Wiley & Sons, Chichester, England, 1982; Watt, J. M., and M. G. Breyer-Bandijk, *Medicinal and Poisonous Plants of Southern and Eastern Africa,* 2d ed., E. & S. Livingston, Ltd., Edinburgh, 1962.

Sources for the information on *Bufo marinus* came from: Abel, J. J., and David I. Macht, The poisons of the tropical toad *Bufo aqua, Journal of the American Medical Association* 56 (1911): 1531–36; Chen, K. K., and H. Jensen, A pharmacognostic study of ch'an su, the dried venom of the Chinese toad, *Journal of the American Pharmaceutical Association* 23 (1929): 244–51; Fabing, H. S., Intravenous bufotenine injection in the human being, *Science* 123 (1956): 886–87; Flier, J., M. Edwards, J. W. Daly, C. Myers, Widespread occurrence in frogs and toads of skin compounds interacting with the ouabain site of Na^+, K^+, ATPase, *Science* 208 (1980):

503–5; Kennedy, A. B., Ecce Bufo: the toad in nature and Olmec iconography, *Current Anthropology* 23, no. 3 (1982): 273–90.

Reports of *Bufo marinus* as a possible hallucinogen include: Dobkin de Rios, M., The influence of psychotropic flora and fauna on the Maya religion, *Current Anthropology* 15 (1974): 147–52; Furst, P., Symbolism and psychopharmacology: the toad as earth mother in Indian America, *XII Mesa Redonda—Religión en Mesoamérica*, Mexico: Sociedad Méxicana de Antropología, 1972; Hamblin, N., The magic toads of Cozumel, paper presented at the 44th annual meeting of the Society for American Archaeology, Vancouver, B.C., 1979. See also Kennedy (1982). Tim Knab has described his experiences in an unpublished manuscript: Knab, T., Narcotic use of toad toxins in southern Veracruz, n.d.

The quote from Ian Fleming comes from page 248 of *From Russia with Love*, Berkley Books, New York, 1982.

There is an enormous literature on the puffer fish. Perhaps the best overview is provided in the excellent paper by C. Y. Kao, Tetrodotoxin, saxitoxin and their significance in the study of excitation phenomenon, *Pharmacological Reviews* 18, no. 2 (1966): 997–1049. An invaluable overview and summary of the symptoms of tetrodotoxin poisoning is provided in: Halstead, B. W., *Poisonous and Venomous Marine Animals of the World*, Darwin Press, Princeton, NJ, 1978. My own scientific findings are presented in: Davis, E. W., The ethnobiology of the Haitian zombi, *Journal of Ethnopharmacology* 9, no. 1 (1983): 85–104, and Davis, E. W., Preparation of the Haitian zombi poison, *Botanical Museum Leaflets—Harvard University* 29, no. 2 (1983): 139–49. For a complete list of references see Kao (1966) and Halstead (1978). Some of the most important papers referred to in my 1983 papers are:

Akashi, T. Experiences with fugu poisoning. *Iji Shimbum* 27 (1880): 19–23.

Clavigero, F. J. *The History of (Lower) California*. S. E. Lake and A. A. Gray, eds. Stanford University Press, Stanford, 1937.

Fukada, T. Puffer fish poison and the method of prevention. *Nippon Iji Shimpo* 762 (1937): 1417–21.

———. Violent increase of cases of puffer poisoning. *Clinics and Studies* 29, no. 2 (1951): 1762.

Fukada, T., and I. Tani. Records of puffer poisonings. Report 1. *Kyusha University Medical News* 11, no. 1 (1937): 7–13.

———. Records of puffer poisonings. Report 2. *Iji Eisei* 7, no. 26 (1937): 905–7.

———. Records of puffer poisonings. Report 3. *Nippon Igaku Oyobi Kenko Hoken* 3258 (1941): 7–13.

Halstead, B. W., and N. C. Bunker. The effects of commercial canning process upon puffer poisoning. *California Fish and Game* 39, no. 2 (1953): 219–28.

Hashimoto, Y. *Marine Toxins and Other Bioactive Marine Metabolites.* Japanese Scientific Societies Press, Tokyo, 1979.

Kawakubo, Y., and K. Kikuchi. Testing fish poisons on animals and report on a human case of fish poisoning in the South Seas. *Kaigun Igakukai Zasshi* 31, no. 8 (1942): 30–34.

Mosher, H. S., F. A. Fuhrman, H. D. Buchwald, H. G. Fischer. Tarichatoxin-Tetrodotoxin: a potent neurotoxin. *Science* 144 (1964): 1100–1110.

Noniyama, S. The pharmacological study of puffer poison. *Nippon Yakubutsugaku Zasshi* 35, no. 4 (1942): 458–96.

Tani, I. Seasonal changes and individual differences of puffer poison. *Nippon Yakubutsugaku Zasshi* 29, nos. 1–2 (1940): 1–3.

Yano, I. The pharmacological study of tetrodotoxin. *Fukuoka Med. Coll.* 30, no. 9 (1937): 1669–1704.

An informative popular article on the Japanese puffer fish, written by Noel Vietmeyer, appeared recently in *National Geographic* 163, no. 2 (August 1984): 260–70.

8. Voodoo Death

The Victorian concerns with premature burial, the protective measures they took, and a marvelous description and sketch of Count Karnice-Karnicki's invention are outlined in: Hadwen, W. R., *Premature Burial—And How It May Be Prevented*, Swan Sonnenschein & Co., Ltd., London, 1905. A number of historical incidences involving premature burial are mentioned in MacKay, G. E., Premature burials, *The Popular Science Monthly* 16, no. 19 (1880): 389–97. Other publications of the era were Fletcher, M. R., *One Thousand Buried Alive by Their Best Friends*, 1890, and Hartmann, F., *Buried Alive*, 1895, both of which were published in Boston. The Townsend case is discussed in Kastenbaum, R., and R. Aisenberg (1972) and mentioned in Watson, L. (1974).

The anthropological literature on voodoo death includes:

Cannon, W. B. Voodoo death. *American Anthropologist* 44 (1942): 169–81.

Cawte, J. Voodoo death and dehydration. *American Anthropologist* 83 (1983): 420–42.

Clune, F. J. A comment on voodoo deaths. *American Anthropologist* 75 (1973): 312.

Eastwell, H. D. Voodoo death and the mechanism for dispatch of the dying in East Arnhem, Australia. *American Anthropologist* 84 (1982): 5–18.

Glascock, A. P. Death-hastening behavior: an explanation of Eastwell's thesis. *American Anthropologist* 85 (1983): 417–20.

Lester, D. Voodoo death: some new thoughts on an old phenomenon. *American Anthropologist* 74, no. 3 (1972): 386–90.

Lex, B. W. Voodoo death: new thoughts on an old explanation. *American Anthropologist* 76, no. 4 (1974): 818–23.

The medical cases of patients surviving near-death experiences (NDE) came from Sabom, M. B., *Recollections of Death—A Medical Investigation*,

Simon and Schuster, New York, 1982. The phenomenon is also discussed in Kubler-Ross, E., *On Death and Dying*, Macmillan, New York, 1969.

9. IN SUMMER THE PILGRIMS WALK

The biogenesis of tetrodotoxin is discussed in Halstead, B. W. (1978). The use of *Duboisia myoporoides* as an antidote to ciguatera poisoning is documented in Dufra, E., G. Loison, B. Holmstedt, *Duboisia myoporoides: native antidote against ciguatera poisoning, Toxicon* 14 (1976): 55–64.

10. THE SERPENT AND THE RAINBOW

An account of the apparition of the Virgin Mary is given in Herskovits, M. J. (1937).

For ethnological information on possession see: Bourguignon, E., *Possession*, Chandler & Sharp Publishers, Inc., San Francisco, 1976; Zuesse, E. V., *Ritual Cosmos—The Sanctification of Life in African Religions*, Ohio University Press, Athens, OH, 1980. For traditional medical studies see: Dorsainvil, J. C., *Vodou et névroses*, Imprimerie La Presse, Port-au-Prince, 1931; Mars, L. *The Crisis of Possession in Voodoo*, Reed, Cannon & Johnson Co., Berkeley, 1977; Price-Mars, J. (1983). For descriptions of possession in the context of contemporary ritual see: Deren, M. (1953), and Lowenthal, I., Ritual performance and religious experience: a service for the gods in southern Haiti, *Journal of Anthropological Research* 34, no. 3 (1978): 392–415; Laguerre, M. S. The festival of gods: spirit possession in Haitian voodoo, *Freeing the Spirit* 5, no. 2 (1977): 23–35.

For discussions of the nature of the vodoun soul see: Métraux, A., The concept of soul in Haitian Vodu, *Southwestern Journal of Anthropology* 2 (1946): 84–92; and Deren, M. (1953). My findings are at odds with both these sources in certain ways, but the most important difference is actually one of semantics. What I refer to as the "ti bon ange" for example, other investigators have called the "gros bon ange"; the actual functions of the two aspects of the vodoun soul are consistent. Métraux himself suggested that research into the nature of the vodoun soul is made "difficult by the wide range of beliefs and theories found among Haitian Vodu worshippers according to their intellectual sophistication, their religious background and contacts with the modern world" (Métraux, 1946, p. 84). Métraux based his interpretation largely on interviews of a single informant at the Bureau of Ethnology in Port-au-Prince. He notes that many of her statements were contradicted outright by the one houngan he also interviewed. Métraux based his conclusions on his "impression that [his informant's] candid statements reflected more closely the general beliefs of the Haitian peasantry" (Métraux, 1946, p. 85). When I commented on the range of professional interpretations by anthropologists to one of my informants I was impressed by his response. He suggested that the diverse opinions reflected the anthropologists' implicit assumption that every vodoun initiate or even every houngan necessarily had the answers to all complex theological questions. Would one expect, he asked, that every French peasant or parish priest would be able to or even be interested in addressing theological issues normally considered by the Vatican alone?

For papers on vodoun ethnomedicine see: Métraux, A., Médecine et Vodou en Haiti, *Acta Tropica* 10, no. 1 (1953): 28–68; Delbeau, J. C., "La Médicine Populaire en Haiti," doctoral dissertation, Université de Bordeaux, 1969; Denis, L., Médecine populaire, *Bulletin du Bureau d'Ethnologie d'Haiti* 4, no. 29 (1963): 37–39.

Notes on Haitian ethnobotany are provided in: Brutus, et al. (1960); Leon, R., *Phytothérapie Haïtienne, nos simples.* Imprimerie de l'Etat, Port-au-Prince, 1959; Les P. Missionaires du T. S. Rédempteur, *Haiti flore Médicinale,* Monastère Saint Gerard, Port-au-Prince, 1943.

A comparison between Western scientific thinking and traditional ways of thought in Africa is discussed in Horton, R., African traditional thought and western science, *Africa* 37, no. 1 (1967): 50–71, 155–87. The entire concept of balance and equilibrium as a key to health in our society is given an excellent treatment by Weil, A. T. (1983).

11. TELL MY HORSE

In addition to the historical sources already noted, *marronnage* is discussed in: Debbash, Y., Le Marronnage: Essai sur la désertion de l'esclave Antillais, *L'Année Sociologique* 3 (1961): 1–112, 117–95; Débien, G., Le Marronnage aux Antilles Françaises au XVIIIᵉ siècle, *Caribbean Studies* 6, no. 3 (1966): 3–44; Manigat, L. F., The relationship between Marronnage and slave revolts and revolution in St. Domingue-Haiti, in Rubin, V., and A. Tuden, eds., *Comparative Perspectives on Slavery in New World Plantation Societies,* The New York Academy of Sciences, New York, 1977; Price, R., ed., *Maroon Societies: Rebel Slave Communities in the Americas,* The Johns Hopkins University Press, Baltimore, 1979. The single most valuable source for my work was Fouchard, J. (1981).

There is an extensive literature on secret societies. Among the general sources are: MacKenzie, N., *Secret Societies,* Collier Books, New York, 1967; Mak, L. F., *The Sociology of Secret Societies,* Oxford University Press, Kuala Lumpur, 1981; Webster, H., *Primitive Secret Societies,* Octagon Books, New York, 1968. A theoretical discussion of the nature of secret societies is presented in Wolff, K., ed., *The Sociology of Georg Simmel,* The Free Press, Glencoe, IL, 1950.

Among the valuable references on the secret societies of West Africa are:
Agiri, B. The Ogboni among the Oyo-Yoruba. *Lagos Notes and Records* 3, no. 11 (1972): 50–59.

Bascom, W. "Secret Societies, Religious Cult Groups and Kinship Units Among the West African Yoruba." Ph.D. dissertation, Northwestern University, Evanston, IL, 1939.

Bellman, B. *Village of Curers and Assassins.* Mouton, Paris, 1975.

Butt-Thompson, F. W. *West African Secret Societies.* H. F. & G. Witherby, London, 1929.

Harley, G. W. *Notes on the Poro in Liberia.* Peabody Museum Papers 19, no. 2, Harvard University, Cambridge, 1944.

Harris, W. T., and H. Sawyer. *The Springs of Mende Belief and Conduct.* Sierra Leone University Press, Freetown, 1968.

Herskovits, M. J. *Dahomey.* 2 vols. J. J. Augustin, New York, 1938.

Jedrej, M. C. Structural aspects of a West African secret society. *Journal of Anthropological Research* 32, no. 3 (1976): 234-45.

———. Medicine, fetish and secret society in a West African culture. *Africa* 46, no. 3 (1976): 247-57.

Little, K. The Poro as an arbiter of culture. *African Studies* 7, no. 1 (1948): 1-15.

———. The role of the secret society in cultural specialisation. *American Anthropologist* 51 (1949): 199-212.

———. *The Mende of Sierra Leone.* London, 1951.

———. Political function of the Poro, pt. 1. *Africa* 35, no. 4 (1965), 349-65.

———. Political function of the Poro, pt. 2. *Africa* 36, no. 1 (1966): 62-71.

Magid, A. Political traditionalism in Nigeria: a case study of secret societies and dance groups in local government. *Africa* 42, no. 4 (1972), 289-304.

Morton-Williams, P. The Yoruba Ogboni cult in Oyo. *Africa* 30, no. 4 (1980): 362-74.

Murphy, W. Secret knowledge as property and power in Kpelle society: elders vs. youth. *Africa* 50 (1980): 193-207.

Ottenberg, S. *Leadership and Authority in an African Society.* University of Washington Press, 1971.

———, ed. *African Religious Groups and Beliefs.* Archana Publications, 1982.

Weckman, G. Primitive secret societies as religious organizations. *International Review for the History of Religions* 17 (1970): 83-94.

For an excellent review article on the African ordeal poisons see: Robb, G. L., The ordeal poisons of Madagascar and Africa, *Botanical Museum Leaflets—Harvard University* 17, no. 10 (1957): 265-316. See also review in Holmstedt, B. (1972).

Zora Hurston's account of her research in Haiti was published as *Tell My Horse*, first in 1938 in Philadelphia. It was reprinted in 1981 by Turtle Island, Berkeley. Her autobiography, edited by Robert Hemenway, is *Dust Tracks on the Road*, and a second edition came out in 1984, published by University of Illinois Press, Urbana, IL.

Typical of the perfunctory dismissals of the possible existence of zombis are: Bourguignon, E., The persistence of folk belief: some notes on cannibalism and zombies in Haiti, *Journal of American Folklore* 72, no.

283 (1959): 36–47; Laroche, M., The myth of zombi, in Smith, R., ed., *Exile and Tradition*, 44–61, Longman & Dalhousie University Press, 1976; Mars, L. P., The story of zombi in Haiti, *Man* 45, no. 22 (1945): 38–40.

The sentiment of these articles was prompted in part by the slew of popular travel books, among them: Craige, J. H., *Cannibal Cousins*, Minton, Balch & Co., New York, 1934, and by the same author in 1933, *Black Bagdad*, Minton, Balch & Co., New York. The first in this tradition was St. John, S., *Hayti: Or the Black Republic*, Frank Cass, London, 1971 [1884]. Others included Seabrook (1929) and Wirkus, F., and T. Dudley, *Le Roi blanc de la Gonave*, Payot, Paris, 1932.

The Haitian secret societies are mentioned in Métraux, A. (1972), Courlander, H. (1960), Herskovits, M. (1937). The traditional view of the Bizango as a diabolical force is presented in Kerboul, J., *Le Vaudou: Magie ou réligion*, Editions Robert Laffont, Paris, 1973. For a different view see: Hurbon, L., Sorcellerie et pouvoir en Haiti, *Archives des sciences sociales des réligions* 48, no. 1 (1979): 43–52. Michel Laguerre's analysis appeared in 1980 in his article, Bizango: a voodoo secret society in Haiti, in Tefft, S. K., ed., *Secrecy*, Human Sciences Press, New York. A discussion of informal justice in rural Haiti appears in Montalvo-Despeignes, J., *Le Droit Informal Haïtien*, Presses Universitaires de France, Paris, 1976.

The reference to the zombi powder that appears in the Old Penal Code [Article 249, cited in Leyburn, J. (1941), p. 164] reads:

Also to be considered as attempted murder, the use that may be made against any person of substances, which, without causing actual death, produce a lethargic coma more or less prolonged. If, after the administering of such substances, the person has been buried, the act shall be considered murder no matter what result follows.

12. DANCING IN THE LION'S JAW

The paradoxical position of the Haitian chef de section is presented by Murray in his 1977 dissertation. For other discussions of the role of the chef de section see: Comhaire, J., The Haitian "Chef de Section," *American Anthropologist* 57 (1955): 620–24; Lahav, P., The chef de section: structure and functions of Haiti's basic administrative institution, in Mintz, S., ed., *Working Papers in Haitian Society and Culture* 4 (1975): 5–81, Antilles Research Program, Yale University.

13. SWEET AS HONEY, BITTER AS BILE

For an examination of land tenure see once again Murray's analysis in his dissertation. A summary of his conclusions is presented in: Murray, G. F., Population pressure, land tenure and voodoo: the economics of Haitian peasant ritual, in *Beyond the Myths of Culture: Essays in Cultural Materialism*, Academic Press, 1980. Issues of land tenure and inheritance relevant to the case of Clairvius Narcisse are discussed in Comhaire, J. (1955), and Underwood, F. W., Land and its manipulation among the Haitian peasantry, in Goodenough, W., ed., *Explorations in Cultural Anthropology* (1964) 469–82.

For the early philosophical views of François Duvalier see: Denis, L., and F. Duvalier, La civilisation haïtienne: notre mentalité est-elle africaine ou gallo-latine? *Revue Anthropologique* 10–12 (1936): 353–73; and their later paper published in 1944, L'évolution stadiale du vodou, *Bulletin du Bureau d'Ethnologie* 2, no. 12: 1–29. Lorimer Denis and Duvalier also published a number of descriptive ethnographic papers based on their observations of various vodoun rituals.

The connection between vodoun and politics in recent Haitian history is examined in: Bastien, R., Vodoun and politics in Haiti, in *Religion and Politics in Haiti*, Institute for Cross-Cultural Research, Washington, D.C., 1966; Nichols, D., Politics and religion in Haiti, *Canadian Journal of Political Science* 3, no. 3 (1970): 400–414; Laguerre, M., Voodoo as religious and political ideology, *Freeing the Spirit* 3, no. 1 (1974): 23–28; Laguerre, M., Voodoo and politics in contemporary Haiti, paper presented at the conference "New Perspectives on Caribbean Studies: Toward the Twenty-first Century," Research Institute for the Study of Man and the City University of New York, August 28–September 1, 1984.

Two extremely negative interpretations of the Duvalier years are: Diederich, B., and A. Burt, *Papa Doc: The Truth about Haiti Today*, McGraw-Hill, New York, 1969, and Rotberg, R., *Haiti: The Politics of Squalor*, Houghton Mifflin Co., Boston, 1971.

Acknowledgments

My research on the zombi phenomenon was undertaken while I was supported by the Social Science and Humanities Research Council of Canada (Doctoral Fellowship). Direct financial support was generously provided by the International Psychiatric Research Foundation, the Wenner-Gren Foundation for Anthropological Research (grant-in-aid 4554) and the National Science Foundation (Doctoral Dissertation Improvement Grant BSN-8411741). My botanical determinations were verified by Professor R. A. Howard of the Arnold Arboretum, Harvard University. Zoological determinations were furnished by the staff of the Museum of Comparative Zoology, Harvard University. Valuable bibliographical material was provided by Professor Bo Holmstedt (Karolinska Institute, Stockholm); Dr. Bruce Halstead (World Life Research Institute); Professor C. Y. Kao (Down State Medical Center, Brooklyn); Professor M. G. Smith (Yale University); and Professor R. E. Schultes (Harvard University). I would especially like to thank Professors Smith and Schultes for their intellectual contributions and encouragement. My understanding of Haitian history is deeply indebted to the excellent work of Dr. Gerald Murray (University of Massachusetts, Boston), Dr. Michel Laguerre (University of California, Berkeley) and Jean Fouchard. The preliminary laboratory work was done by Professor Leon Roizin of Columbia University. For other laboratory work and advice I am indebted to Dr. Laurent Rivier (Université de Lausanne) and Professor James Cottrell and Dr. John Hartung (Down State Medical Center). The Zombi Project was born of the vision of three men: Mr. David Merrick, Professor Heinz Lehman, and the late Dr. Nathan S. Kline.

The work in Haiti was made possible by the cooperation and active support of many individuals. Dr. Lamarque Douyon shared his insights concerning medical aspects of zombification and introduced me to Clairvius Narcisse. Medical records were kindly furnished by the staff of the Albert Schweitzer Hospital. In Port-au-Prince I was kindly received and assisted by Leilas Desquiron, Dr. Max Paul, and his staff at the Institut National Haitien de la Culture et des Arts, Eleanor Snare at the Institut Haitiano-Americain, and Lesly Conde at the National Office of Tourism. In rural Haiti I worked directly with several people who openly shared their knowledge. In particular I would like to thank Jean Baptiste, Jacques Belfort and Madame Jacques, Michel Bonnet, Andrés Cajuste, Andrés Celestin, Robert Erie and his wife, Carmine, Ives François, César Ferdinand, Jean-Jacques Leophin, La Bonté, Miriatel, Jean Price-Mars, Solvis Silvaise, and Marcel Pierre. All of these individuals, some of whom are herein identified by pseudonyms, were directly responsible for the success of the project. Finally I would like to acknowledge Herard Simon and Max Beauvoir. Herard Simon and his wife, Hélène, are serviteurs of the most profound awareness. A great houngan, Herard offered his spiritual and physical protection without which this project would never have been completed. Max Beauvoir was also directly responsible for the success of the project. He and his wife, Elizabeth, and his daughters, Rachel and Estelle, offered me their home and generous hospitality as well as their total support at the most critical moments. Rachel worked with me on every phase of the fieldwork and her enthusiasm, courage, and generosity were unfailing. My debt to her, as to all the people of Haiti who received me so kindly, should be readily apparent in the text of the book.

In Virginia, Lavinia Currier gave me a place to sit still long enough to complete the manuscript, portions of which were reviewed by her and Charlie Fisher. Monique Giausserand read the entire manuscript and her comments and advice at each stage of its preparation were invaluable. Raymond Chavez and Harmon and Virginia Stevens kept my spirits up while I was writing, and for technical assistance I am indebted to Timothy Plowman and Penny Matekaitis. My agent, Jane Gelfman, supported the book from the start and introduced me to my editor, Don Hutter, without whose interest and patience this book would not have been completed. Finally, I would like to thank Monique for support and constant love.

WADE DAVIS
The Plains, Virginia

Index

About the Author

Wade Davis holds degrees from Harvard University in Anthropology and Biology, and recently received his Ph.D. in Ethnobotany. Mostly through the Harvard Botanical Museum, he has worked in the field as a plant explorer, ethnobotanist, and photographer, investigating fifteen tribal groups in eight Latin American nations, before undertaking the assignment to Haiti that led to his writing *The Serpent and the Rainbow*. A native of British Columbia, Mr. Davis has worked as a logger, surveyor, hunting guide, and park ranger. He has published numerous scientific papers and lectured extensively.